The
Mariner's Guide to
OCEANOGRAPHY

The Mariner's Guide to OCEANO

Edited by Nixon Griffis

Line drawings by Doreen Szczupiel

GRAPHY

Hearst Marine Books, New York, NY

Library of Congress Cataloging in Publication Data

Main entry under title:

The Mariner's guide to oceanography.

1. Oceanography—Popular works. I. Griffis, Nixon.
GC21.M3 1984 551.46 84-10869
ISBN 0-688-03976-6

Printed in the United States of America

First Edition

1 2 3 4 5 6 7 8 9 10

BOOK DESIGN BY MARIA EPES

PREFACE

A sometime definition of a boat is "a hole in the ocean surrounded by wood into which you pour money." What anathema! Better it should read "a structure of beauty with which you gain a whole spectrum of inspiration from the surrounding seas."

Water-men are luckier than most because man is a land animal, and going to sea still connotes a spirit of adventurism and intellectual curiosity in a relatively unknown realm. Marine science is progressing on a geometric scale, but the vaster body of knowledge has only been garnered in the last fifty years of this century. A famous marine biologist once said, "Twenty years ago we did not know the complete life cycle of a single species of any of our important marine food fish." Today we know most of them.

This book is written to aid the reader in his understanding of the water world and his appreciation of the fantastic multitude of life forms that exist there; also of the geology of the shoreline, the vicissitudes of waves and weather, the fragility of the ocean environment, and the relics on the ocean floor of man's mistakes in navigation. It is intended as an amateur's overview, not as a scientific treatise. The reader will not come away expecting to know the scientific names of fish; in fact, one head of a marine laboratory once said of his research-vessel captain, "If he says *clupea* once more instead of herring, I am going to hit him right in the head."

As with life on land, in the sea nature attempts to fill every niche and cranny. A bare artificial tire reef lowered to the ocean bottom will reach the biological equivalent of a climax forest within three to five years depending on the temperature of the water. Temperature for some life forms ranges from the frigid waters of the Arctic and Antarctic to the tropical seas, and yet some fish must exist with a temperature gradient of only five degrees. Pressure, which is an almost imperceptible force on land animals, plays an exceptionally important role on sea life. Many fish cannot be brought up alive for aquarium exhibits because the change from high-pressure to low-pressure environments explodes them internally. Many marine animals conduct their entire lives in total darkness, and for this reason luminosity occurs far more in sea creatures than in land animals. There is only a handful of insects at sea, but land species number more than 80,000. However, the spider clan is represented by the lowly horseshoe crab and the bizarre-looking pycnogonids, or sea spiders, existing throughout the oceans. The total number of animals and plants in the underwater world would be meaningless because of its size and variation within season, but of species there are over 5,000 classified seaweeds, 20,000 fish, 35,000 crustaceans, and 500,000 invertebrates.

Reading this book is only the beginning of the adventure. See from your boat the glory of birds in the wind, the rare glimpse of a shy sea mammal, the diverse patterns of algae under the microscope, the delicate colors of the seaweeds, the fish and the patterns they make on the surface, the range of the shoreline and its geological formations, the variations of worms and other creepy-crawlies from docks and pilings, and find unlimited enjoyment and knowledge from the world of the sea.

NIXON GRIFFIS
The Griffis Foundation
New York, New York

Foreword: The NEW YORK AQUARIUM and the OSBORN LABORATORIES of MARINE SCIENCES

What's going on below the ocean's surface? As the environmental consciousness of the public has grown in recent years people have recognized that the water world is filled with mysteries, with wondrous creatures and discoveries yet to be made. Fishermen, just by doing what they do, get more of a glimpse than most of us. Occasionally they even haul up some never-before-seen oddity. The relatively small segment of the population seriously devoted to diving gets a firsthand look that most of us will never share. Only a mere handful of individuals—

scientists mostly—will ever personally probe the ocean depths.

The mysteries, and their inaccessibility to most, are what make the New York Aquarium such a popular diversion. The Aquarium is a place of secrets revealed. A place where visitors—without traveling far, without donning scuba gear and entering a foreign environment—can safely peer inside the famous (some would say infamous) Bermuda Triangle . . . explore the kaleidoscopic patterns of community life in a coral reef . . . or wend a lazy path through the tangle of a mangrove swamp.

The Aquarium's roots trace back to its opening as a municipal institution, late in 1896, the first public aquarium in the United States. Located in Battery Park at the southern tip of Manhattan, the original aquarium became in 1902 a division of the New York Zoological Society, which also operates the Bronx Zoo and an international conservation program. When Manhattan was linked to Brooklyn by a tunnel at the Battery, construction forced the Aquarium to close and plans were drawn for a new facility. World War II intervened, but finally, in June 1957, a modern New York Aquarium opened on the beach at Brooklyn's Coney Island.

Fish of all shapes, sizes, colors, and behaviors predominate but visitors can also get an intimate glimpse into the lives of invertebrates like the octopus and flowerlike anemones, as well as such aquatic mammals as the seal—a panorama of life ranging from some of our planet's smallest creatures to some of the largest. If any whale can be considered cuddly it is the plumpish, ever-smiling beluga. Audiences flock to see them—and their even more athletic cousins the dolphins—leaping and splashing through behavioral demonstrations.

Since its opening on Coney Island nearly thirty years ago the New York Aquarium has never ceased evolving, growing, improving. Recent additions include the Native Sea Life Building, in which Northeasterners can, often for the first time, learn what lurks in the waters just offshore, and a giant shark tank where visitors can succumb to the *Jaws* syndrome, respectfully wide-eyed at the streamlined menace of some of the world's most efficient predators.

More is on the way: a new arena (with above- and below-water viewing) for the belugas and dolphins to please their human admirers; the Discovery Cove to involve family groups in participatory learning about the marine environment; and a rocky coast habitat for seals, penguins, and sea otters. More secrets to be shared.

One component in the Aquarium complex remains unseen by the general public even though the work inside touches their lives, often deeply. It is strictly a research institution: the Osborn Laboratories of Marine Sciences, opened in 1967. Among the enormously varied programs within this enterprise, scientists probe marine organisms for new antibiotics, fungicides, and other pharmaceutical products . . . search for clues to alleviate human cancer, blood disorders, and other diseases . . . and attempt to identify methods of improving yields in aquaculture that will, ultimately, help to feed a hungry world. Royalties from *The Mariner's Guide to Oceanography* will aid in projects such as these.

That brief résumé describes but a fraction of the long-term research, past and present, inside the Osborn Laboratories. This, too, is a place of mysteries—the unsolved kind—and, as the Osborn scientists have proved frequently over the years, today's mysteries are the revelations of tomorrow.

—EUGENE J. WALTER, JR.
Editor in Chief, *Animal Kingdom* magazine
New York Zoological Society

CONTENTS

PRESERVING the SEA

Senator Lowell Weicker, Jr.

The other evening I was engaged in a favorite occupation—relaxing with my two boys, Sonny and Tre, ages four and three. We were watching a favorite movie of theirs—Jules Verne's classic *Twenty Thousand Leagues Under the Sea.* It wasn't long before other friends and family drifted in. I was surprised at their admiration of Verne's foresight when it came to submarines and scuba gear, and yet their skepticism of Verne's discussions of farming the ocean floor and seeing Captain Nemo live totally off the bounty of the sea.

With the exception of a nod to Jacques Cousteau for having brought the beauty of the oceans to our attention, it was as if nothing had occurred since the days of Jules Verne to make those oceans an integral part of our lives. Even more to the point of this chapter, it was as though the sea floor itself was fiction rather than a reality with monumental practical significance for the future of the world.

The time will come in America when the oceans are recognized for their vast wealth of renewable food and energy. When that recognition comes, will it be accompanied by an equally enthusiastic commitment to a conservation ethic necessary to make the resource last for eternity? Or will it simply be used to satisfy the wants of one generation? As a nation, we have a reputation for gobbling up whatever strikes our fancy and then leaving behind a wasteland. That must not happen to the oceans, the last great "on planet" resource. Thus the attempt by this voice to encourage his fellow Americans to understand their watery future.

In 1975, I had finished two years of Watergate investigation, and wanted to put that essentially negative exercise behind me by creating something useful with the power given a United States Senator by his constituents. Happenstance led me to visit an ongoing marine experiment in the Bahamas called Operation Score. Transported three miles offshore to waters that were approximately sixty-five feet deep, I could barely see the outlines of a large stationary cylinder with people swimming in and out of it. I was told that a group of scientists was living in that cylinder and using it as a base from which to conduct a variety of experiments on the ocean floor.

The scientists were using a technique known as saturation diving, so named because one's body tissues are "saturated" with the inert gases for the particular depth he is living at. The advantage of being "saturated" and living on the ocean floor is that the marine scientist can remain on the bottom, conducting experiments, for extended periods of time without worrying about "the bends." At the completion of his stay on the bottom, whether it be one week or one month, the decompression requirements before returning to the surface remain the same. The development of saturation diving as a research tool for the marine scientist has opened up an entire range of research objectives that were impossible from a surface diving orientation.

Well, curiosity got the cat, and twenty-four hours and one crash scuba course later, I was sixty-five feet down, looking at the world of Jules Verne in real life. Two young scientists had a complete garden laid out row by row. Strings delineated the different species. It looked exactly like my victory garden, Long Island, New York, circa 1942. Farther away, a school of grunts was being herded in an underwater pen much like sheep corralled on land. I was viewing some basic experiments in aquaculture, an exciting field that has the potential for solving world food problems. Other people were taking sediment samples, measuring currents, getting rock samples, examining the coral for diseases, and so on. The scene was one of people doing normal work—but beneath the ocean surface. And, yes, what had appeared topside to be a cylinder was just that. Resting on four legs, the sixteen-foot-by-eight-foot cylinder housed bunks, cooking equipment, marine band radio gear, etc. It was a stationary submarine from which scientists issued forth at will to do their underwater work.

From that moment on, the oceans became part of my world. They didn't belong just to Captain Nemo, Jacques Cousteau, Herman Melville, Lloyd Bridges, or deep research submersibles such as *Alvin.* They weren't a freak or entertainment. They contained an order, a logic, a beauty, a quiet as magnificent as anything to be seen on earth.

Eight years have gone by. The Hydrolab, originated and kept running for so many years by the Perry Foundation, is now owned by the U.S. government. It is now located off the island of St. Croix and run by Fairleigh Dickenson University. Its interior is the same in size, but much improved in terms of equipment and conditions. Air conditioning, fresh-water showers, nourishing hot food, wall-to-wall carpeting, and an experienced topside crew augment the operation.

Scientists from all over the world have called it home for periods of up to eight days as they've learned firsthand how the ocean looks, moves, and lives. These men and women find their lives changed forever; they know the inevitability of life beneath the sea. And they also understand that, like the land, it is a fragile, balanced environment. They want no "Silent Springs" here. I've been a privileged part of this group, having, on four different occasions, lived in the habitat for periods ranging from three to six days. Along with beauty, I've seen the beginnings of man's destructiveness.

Coral, the fragile cornerstone of the tropical reef ecosystem, can be buried and killed by sediment washing off the land as commercial and residential castles are built in the air. "Ghost pots," abandoned fish traps and lobster pots disconnected from their surface floats, are lethal havens for inquisitive fish and crustaceans; they continue to trap and kill for years. The exquisitely delicate and beautiful black coral is hacked from its base, to be polished and sold to tourists as souvenirs.

This coral is found only rarely and at great depths in areas where it was previously abundant. The marine fishery resource is harvested with little thought given to managing the stocks. And so even now there are choices to be made.

Today, because of sensitive administrators of the National Oceanic and Atmospheric Administration (NOAA) like Dick Frank and John Byrne, scientists, divers, officials such as Bob Wicklund, Bob Dill, John Ogden, Sylvia Earle, and Kevin McCarthy, and artists such as Stan Waterman and Al Giddings, we stand a chance of making our entry into the oceans an occasion of pride and achievement, rather than an exercise in environmental gluttony or an invitation to international conflict. The solution is not simply to throw money at the problem. In the long haul ahead, the patience and sustained commitment of the American people *are* going to be needed.

First, an oceans policy has to be devised. The questions needing answers are: What are our national objectives, our international objectives, our environmental concerns, and our resource concerns? What is the role of the federal government? From past experience I believe it will have to be substantial. Private companies have cooled on their ocean commitments once the bottom line showed up in red ink.

Second, NOAA should be a separate entity, not under the Department of Commerce, not under Treasury, not under Interior. We might contemplate merging NOAA with the Coast Guard. In any event, a well-ordered use of the oceans is mainly a scientific venture, thus expensive and unpopular, and should stand on its own and not compete with the politically more attractive portions of other departmental budgets.

Third, the entire U.S. academic community should be involved, and that necessitates an acceleration of funding to the Sea Grant Program. Only if all disciplines are involved will our knowledge of the sea be of the breadth and depth that mastery of any subject matter demands.

Lastly, men and women will learn about the water only by being in it. Manned underwater research has proven basically feasible and now deserves the most sophisticated approach and technology. No laboratory, no extrapolation, no model, no educated speculation substitutes for the human mind in the actual environment it seeks to understand and control. In other words, to my friends in the ocean scientific community—get wet!

These thoughts don't presuppose a race against the Russians, nor are they the science-fiction ravings of an excessively imaginative United States Senator. They argue for an informed, common-sense response to the most exciting opportunity of our times. They are a plea to save from destruction that about which we know nothing. Man has always meant to walk hand in hand with nature, not against it, nor away from it. The next time you stand at a lake's edge or see a river flowing by, or sail over the oceans and bays, ask yourself what you know beyond what you see. Today's answer won't suffice for our future.

SUGGESTED READING

Under the Sea Wind: A Naturalist's Picture of Ocean Life, by Rachel Carson (New York: Oxford University Press, 1952).

Protected Ocean: How to Keep the Seas Alive, by Wesley Marx (New York: Coward, McCann & Geoghegan, 1972).

Senator Lowell Weicker was born in 1931 in Paris, France. He received degrees from Yale University and the University of Virginia School of Law. He served in the U.S. Army as a first lieutenant from 1953 to 1955. Prior to entering public service, Senator Weicker practiced law in Greenwich, Connecticut.

Senator Weicker has never lost an election for public office. He served as state representative (1962 to 1968) and first selectman of Greenwich (1963 to 1967), prior to becoming the U.S. representative from

Connecticut's fourth congressional district in 1969. In 1970, after just one term in the House, he was elected to the Senate. He was returned to the Senate in 1976, earning the largest number of votes ever in the history of Connecticut politics, and again in 1982.

Senator Weicker is chairman of the Small Business Committee, which oversees government activities in the vital sector of the nation's economy. He serves on the Appropriations Committee and is chairman of the Labor, Health and Human Services, Education Subcommittee. He is a member of the Labor and Human Resources Committee and chairs its Subcommittee on the Handicapped. In addition, he sits on the Energy Committee and is chairman of the Energy Conservation and Supply Subcommittee. Senator Weicker has earned a reputation in more than two decades of public service for leadership on issues as diverse as energy conservation, rights of the handicapped, and oceans research.

SUGGESTIONS of a WHALE WATCHER

Dr. Roger Payne

The mammals of the sea are the whales (which include porpoises), sea lions and seals, manatees and dugongs, and the sea otter. Representative species from all of these groups except the dugongs can be seen in coastal waters of the lower forty-eight states. No other country is really as lucky as we. Sea otters can be seen only on the Pacific coast, and manatees only in a few places along the coastlines of the southeastern states (Florida being the best). But elsewhere, marine mammals frequently can be seen. In many popular boating areas they are not really rare; you simply must expect to see them and you will begin to notice them. It is surprising how many people fail to notice marine mammals even in areas where they are abundant. I have talked with people getting off boats that I had seen sailing all afternoon among a group of whales I was studying, only to see them crestfallen when I told them whales had been so close by. They had never noticed them.

In the Caribbean I once interviewed about twenty-five professional boat skippers who took tourists on week-long trips all through the winter. Their routes passed right through a humpback whale breeding ground—a ground that has an operating whale fishery along its shores. In most of these cases no one had ever seen a whale. In many cases the skippers assured me that there were no whales present in the area. One man even added, "Believe me, if there were whales around here I would see them—I've cruised these waters for eight years and I have never seen a whale, even though I'm always looking." A local fisherman tied up in the next boat slip over told me that he saw whales in the same waters almost every day in winter. He was afraid of whales, and so he really was looking out for them.

An important thing to remember when searching for marine mammals is that although many species roam over vast areas, most species we know much about prove to be creatures of habit. They return at the same times each year to the same areas for feeding and calving. If you have a friend who saw whales on a cruise through your area last year, find out exactly where and when they were seen, and then go look in the same area at the same time this year and you may very well see what your friend saw. If someone comes in with a report of whales from yesterday or even last week, don't despair because you missed them. What you are really being told is that the whales are now around, and if you go out in the next few days in the same area you have a good chance of seeing them too. The same principles hold true when looking for sea lions, seals, and sea otters. They tend to be concentrated on particular beaches or along specific stretches of coast at quite predictable times each year.

Once you have found a marine mammal, you will probably want to know what species you are looking at. Several excellent field guides exist. You will need one, for most marine mammals are somewhat tricky to identify in the sea despite their large size. Very little shows at the surface when they rise to breathe, and their markings are usually subtle. But take heart; there will probably be only a few species in your usual boating area, so you can quickly become expert in identifying them if you spend a bit of time at it. Then, when the occasional rarity comes by, you will spot it at once.

In my opinion, by far the best field guides for identifying whales, dolphins, and porpoises are the following: for the Atlantic: *Whales, Dolphins and Porpoises of the Western North Atlantic, a Guide to Their Identification* (NOAA Technical Report NMFS CIRC-396). It is by Leatherwood, Caldwell, and Winn, and in spite of the numbers following its title, it is eminently readable. For the Pacific the equivalent guide is: *The*

Mother and infant humpback whales RICHARD ELLIS

Whales, Dolphins and Porpoises of the Eastern North Pacific. A Guide to their Identification in the Water (NUC TP 282). It is by Leatherwood, Evans, and Rice. Both can be purchased by writing the Superintendent of Documents, U.S. Government Printing Office, Washington, D.C., 20402. Good information for identifying seals and sea lions is found in a book by Victor Sheffer called *Sea Lions and Seals of North America.* It can be ordered through bookstores. The volume on mammals in the Peterson Field Guide Series contains good information on all the marine mammals. To identify whales at sea, however, I feel you need more help than that volume affords.

Once you have figured out what the sea mammal is that you are watching, you will probably wish to get closer in order to see what it is doing. If you see something really interesting or get photographs of distinctive markings on its body, you may wish to share your data with scientists working with the species that you saw. Let us take these objectives one at a time.

First, getting closer: The first thing to remember here is that all marine mammals are now protected from harassment by law—a law enforced by stiff fines of up to $10,000! I take this as a sign that our civilization is capable of true progress. I hope you will agree and will respect this excellent law, and will use opportunities that arise to remind others, gently, of its existence. Harassment as defined in the law means, basically, operating your boat in a manner that changes the normal behavior of the mammal you are watching. In practice, this means you may not approach an animal closer than about 300 feet, and you must certainly never pursue it. I can guarantee you that if you do pursue it, you will get far poorer views than via a slow, patient approach. Harassed animals make themselves scarce almost magically. It is amusing to watch the reactions of impatient tourists aboard whale-watch vessels run by professionals. They are frantic for the "deadhead captain" to pay attention and get on over close to that whale before it goes down so that they can get some pictures. Of course, when the whale finally comes up right next to the boat twenty-five minutes later, they never realize that if the captain had not been so patient they would never have seen the whale close at all.

Mother and infant right whales RICHARD ELLIS

Given the law and the difficulty of maneuvering up close to a whale, how *do* you get a close-up look? Many marine mammals are naturally curious, and their normal behavior will include approaching your boat. If you wish to be approached, you must make yourself available for approaching. That means you must motor or sail slowly (avoiding changes in engine speed that may frighten the whale) to within 3 or 400 feet and then stop. But when you do, do not stop your engine. Leave it idling. Whales will, at least in my experience, approach a boat more often when its engine is running. I suspect they like to know where the boat is, and the engine noise informs them of your location at all times. Sneaking up on a whale may panic it, and it will vanish. In fact, the most I ever scared a whale in my life was when I paddled silently up next to it, a right whale, in a kayak. It took off and swam flat out for five miles, leaving the biggest wake I have ever seen a whale make. When it slowed, it circled nervously for about an hour while continuing to move away.

When a whale comes close alongside your boat, don't make loud noises; whales will often flee from loud sounds. And don't worry; a calmly approaching whale won't hurt you, even if it touches your boat.

If you see a whale at close quarters, you will never forget the experience. Even if you once walked on the moon, such a whale sighting will fall within the most memorable experiences of your life. Remember, patience is necessary for a close encounter. If you maneuver in the vicinity of whales all day, always keeping your distance and always moving slowly with long stops in between, you will probably be closely approached at least once. It is really up to the whales as to when you will see them up close.

Whale species can be distinguished, after a sighting, from good photographs. In several cases there are scientists studying hundreds, even thousands, of whales that they recognize as individuals by photographs of their natural markings. Like fingerprints in humans, these markings are distinct indicators for each individual. The tail flukes, raised into the air before diving by some species (particularly the increasingly common humpback whale), are rich in these individual marks. The dorsal fin, or body scars, or white blazes are also useful and should be photographed. If you wish your photographs to be valuable to scientists for identification, use a telephoto lens (300 mm is the biggest you can hold steadily) and don't save film! Take at least a roll of any individual whale, and don't stop taking pictures until you are sure you have a clear set of photographs. The right whale, the rarest whale on earth, yet seen with some regularity in waters north of Long Island in the boating season, can be individually identified by photographs of the strange bumps on the top of its head. The best identification photographs come from high up on the boat. In this case, don't even stop at one roll of film. Many pictures usually are needed to identify an individual right whale.

Scientists are very grateful to receive good copies and do not need your originals. There are several centers where records are kept on individual whales. Steven Katona at the College of the Atlantic in Bar Harbor, Maine, will tell you which individual your North Atlantic humpback whale is (from pictures of its flukes) and where it was seen before. James Darling at the Western Canada Whale Research Center in Victoria, British Columbia, will tell you the same thing in exchange for pictures of the raised flukes of North Pacific humpbacks. Eleanor Dorsey at the Center for Long-Term Research, Lincoln, Massachusetts, is interested in pictures of the dorsal fins and upper bodies of minke whales from the Pacific Northwest. Steven Schwartz at Whale Research Associates, San Diego, California, works with individual gray whales which he identifies from photographs of flukes and back markings. Richard Sears of Mingun Island Research in Quebec keeps track of the St. Lawrence River blue whales. Scott Kraus at the New England Aquarium, Boston, Massachusetts, works with other scientists at the University of Rhode Island and at the Woods Hole Oceanographic Institute to identify the western North Atlantic right whales. Charles Mayo at the Center for Coastal Marine Studies in Provincetown, Massachusetts, collects data on western North Atlantic fin and humpback whales. Kenneth Balcomb of the Moclips Cetological Society, Moclips, Washington, works with killer whales, which he identifies by the same means.

Regional offices of the National Marine Fisheries Service also appreciate photographs and will help get them to interested scientists.

Write to the Platforms of Opportunity Program, National Marine Fisheries Service, Tiburon, California, 94920.

If you find a stranded whale while beachcombing, photograph it from both sides as well as from the front and rear. Note its sex if possible (and confirm with a close-up photograph of the ventral surface showing all openings). Measure its length (not along the animal's curving outline but along the ground in a straight line parallel to the body from the tip of the snout to the notch in the tail, rather than to the end of the longest points of the tail). Pictures taken from a point at right angles to the body axes are of most use. Send the photographs to James Mead (Division of Mammals, U.S. National Museum, Smithsonian Institution, Washington, D.C., 20560) for identification.

A last important note: Always include date and location. Be very specific about where the boat was—don't worry if you were no surer than five miles of your position; we have all been in the same boat, so to speak, but do note your uncertainty.

It is hard to overestimate the value of photographs taken by casual boaters. Your information may very well fill in missing links in migration routes of these species. Scientists who receive well-documented photographs will greatly appreciate them, and sometimes interesting friendships (and perhaps missions in the future) for you may grow out of these contacts.

Now, what are the whales doing? This can be tough to figure out, for scientists no less than for boaters. Some obvious behaviors are breaching (jumping high from the water and splashing back with a monumental cloud of spray), lobtailing (slamming the water's surface repeatedly with the flat of the tail), and flippering (like lobtailing, but the flippers are used). We have evidence for two species that these noisy behaviors are sometimes used as signals between nearby individuals. They may also sometimes represent an indication of an aggressive mood. Until this point can be cleared up, do not closely approach a breaching whale. There are three instances on record of humpback whales crushing small boats that moved very close to the whale when it was breaching.

In summer, if you see whales lunging at the surface singly or in

Breaching sperm whale RICHARD ELLIS

groups, they may be feeding, particularly if birds are circling above them and landing in the water. A curious form of feeding in several species (particularly the humpback whale) involves releasing air underwater either in columns of bubbles or as a single large cloud of air. The whale then comes up next to the bubbles or through them to scoop up a mouthful of food. Watch closely if the mouth is just completing its closure when the whale surfaces. Sometimes you can see the prey it is collecting jumping out of the mouth or frantically darting about within it.

Whales rolling about together with much bodily contact are socializing. Sometimes this is mating or attempted mating. Other behaviors are more subtle, and description of them is beyond the space available here. But keep watching; in most species even very obvious behaviors such as mating have never been clearly described. Neither has calving, or suckling of the young, or the exact details of some of the fancier feeding techniques such as those involving air bubbles. Given the especially favorable conditions under which clarification of these behaviors could be possible, the laws of probability suggest that it is unlikely a professional marine mammalogist will be in the right place at the right moment to see them. It will probably be an afternoon boater out for a spin who has the luck. If you are the one, note down everything you see. Try not to interpret what was going on. Simply say what you saw, exactly, and leave it at that. Write notes immediately after the observation. Every minute counts in getting the details down and straight. The most important things that most observers leave out are the really obvious things. Describe if you can exact times, noting how you measured or estimated them. Note the speeds of swimming involved, distances between animals, their postures, orientation to the surface, to each other, etc. The scientific world will be grateful to you for such observations.

If you get a good chance to watch the whales in your area, you may wish to do something to affect the fate of your newly discovered neighbors. Any of the many conservation organizations in your area is likely to be grateful for your aid, whether of money or time.

Taking some action on behalf of marine mammals is, in my opinion, the most important follow-up to a day of whale watching. If you have seen whales in their wild element from your own boat, you are one of a tiny minority. Every one of our voices counts. If whales cannot rely on our help, if we lucky few don't speak out for them (and return the grace they introduced into our lives), then no one else will, and the opportunity will surely vanish in the future for anyone to see the mammals of the sea.

Dr. Roger Payne, born in New York City in 1935, received his Ph.D. from Cornell University in 1961. After doing postdoctoral work and teaching at Tufts University, Dr. Payne joined the staff of the New York Zoo-

logical Society as a research zoologist. Since then he has been an adjunct associate professor at the Rockefeller University and a research scientist at the World Wildlife Fund, U.S.

His work focuses on the behavior of whales with emphasis on their sounds. It has involved fifty-two expeditions by Dr. Payne and his students to nineteen locations in five oceans. Discoveries by Dr. Payne and members of his laboratory include the fact that humpback whales sing songs which they change according to complex laws and the fact that in deep water the sounds made by finback and blue whales can propagate across entire oceans. He has also made a thirteen-year study of the biology of the southern hemisphere right whale.

His work for conservation of whales includes the release in 1970 of a record, "Songs of the Humpback Whale," which became a best-seller. It was followed in 1977 by a second whale recording, "Deep Voices." A third recording by Dr. Payne was published by the National Geographic society in 1979 in an edition of 10.5 million copies, making it the largest single pressing in the history of the recording industry. The recordings have also inspired numerous works of human music, as well as dance and theater compositions—all of which Dr. Payne encourages in an effort to make whale sounds become a part of human culture.

The TRUTH About SHARKS

Richard Ellis

Most mariners—or non-mariners, for that matter—believe in the Gospel According to Jaws; that is, that all sharks are voracious and malicious man-eaters, cruising the waters of the world with their dorsal fins slicing ominously through the water, waiting for the opportunity to tear anyone in the water limb from limb and, in a froth of the victim's blood, gobble him up.

While it is true that *some* sharks, in *some* circumstances, have indeed attacked people and even killed them, the normal behavior of most sharks is exactly the opposite. For the most part, sharks are timid creatures that would prefer to swim away from any unusual activity rather than investigate it. It's not that sharks are not carnivorous; it's just that most of them are "prey-specific," with teeth and habits designed to eat certain prey species, and neither men (nor women) nor boats are high on the preferred menu of most species.

When the word "shark" is mentioned nowadays, most people think of the great white shark, the star of *Jaws*, and a formidable creature indeed. White sharks (which are not white at all, but rather a dirty gray above and lighter below) have been recorded in most of the world's temperate waters, but they are nowhere common. They can be found in some numbers in those areas where there are proper food sources for them: they feed mostly on seals and sea lions, but they also consume fish, squid, rays, and even other sharks. The white shark is the largest carnivorous fish in the world, but it is not the largest shark, an apparent contradiction that will be dealt with later. The largest white shark accurately measured was a 21-foot specimen, taken in Cuban waters in 1948. It weighed over 7,000 pounds. Since the publication of *Jaws*, fishermen on both coasts of the United States have exercised what can only be described as a vendetta against the great white. There have been 16-footers taken off Montauk, Long Island, and off San Francisco. In South Australia, where the first white sharks were filmed underwater for Peter Gimbel's sensational *Blue Water, White Death*, there is an ongoing sport fishery for white sharks. (The largest fish ever taken on rod and reel was a great white, 16½ feet long, and weighing 2,664 pounds.) It is even possible to hire Rodney Fox, a man who was nearly killed by a white shark in 1963, to take you down in a shark cage to see these awesome creatures, as they feed on chunks of horse-meat.

Although the great white is the quintessential shark, with all the equipment that has made his kind so fearsome (razor-sharp teeth, soulless black eyes, slack-jawed grin, etc.), there are another 300-odd species of sharks, and very few of them could be classified as fearsome or even dangerous. (Any wild animal, of course, from an antelope to a woodchuck, can be dangerous if provoked, but we are discussing situations *without* provocation.) As mentioned earlier, the great white at a maximum length of 21 feet is the largest *carnivorous* shark—that is, feeding on large, warm-blooded prey—but there are at least two species that grow larger and are totally harmless to man. These are the whale shark *(Rhincodon typus)*, and the basking shark *(Cetorhinus maximus)*. The whale shark, which can reach a length of 40 feet and a weight of ten tons, is the largest fish in the world. It is a plankton-eater, and swims placidly through tropical waters with its six-foot mouth agape, ingesting huge quantities of microscopic animals, such as plankton.

In the temperate waters of the Northern Hemisphere, another giant shark can be found, the basking shark. This creature, which is often mistaken for the white shark, is dark gray in color and looks considerably more "sharklike" than the whale shark, since it has a pointed snout and a high, triangular dorsal fin, while the whale shark is a fawn or

Great white shark RICHARD ELLIS

brownish-gray fish, with a series of ridges and white spots, making it easily identifiable. Basking sharks get their name from their habit of loafing at the surface, often in large groups. They can open their mouths to an astonishing degree, and like the whale sharks, they also feed on microscopic organisms.

In the polar waters of both hemispheres, there can be found a large, slow-moving shark, known in the north as the Greenland shark *(Somniosus microcephalus)*, and in the south as *S. antarcticus*. These sharks—which can reach a length of twenty feet—are often found at prodigious depths, but they are almost never seen by seafarers, and should not concern us here.

Another large shark was discovered in 1976, off Kaneohe, Hawaii. A U.S. Navy research vessel was hauling in a parachute sea anchor, when a huge shark was found to have swallowed the orange and white cargo 'chute. Because one of the naval officers did not recognize it as any known species, the 13½-foot carcass was brought to the Waikiki Aquarium, where it was examined by Dr. Leighton R. Taylor, an authority on sharks. He realized that this 1,600-pound creature with the rubbery-lipped mouth was a completely unknown species, and in 1983, the scientific description of *Megachasma pelagios* (nicknamed "Megamouth"), was published. To date, this is the only specimen to have been found, so we have no way of knowing how large they can grow. Because this shark was caught at a depth of 500 feet and, being a fish, has no reason to come to the surface, it was only the remarkable coincidence of the parachute and the feeding shark—which may have mistaken the parachute for a swarm of plankton—that made it possible for us to know of its existence. Because it is a plankton-eater, it would obviously not take a baited hook, and because sharks sink when they die, this deep-water species would not even wash ashore if it died of natural causes.

But for every large species of shark—over six feet, let us say—there are probably twenty-five species that are small and harmless. There are entire families of sharks that never exceed three feet in length, and others that inhabit such deep waters that they are never seen, except by deep-water trawlers or robot cameras.

Tiger shark RICHARD ELLIS

The largest genus of sharks is the Carcharhinidae, which includes most (but not all) of the better-known large sharks. Blues, tigers, bulls, duskies, sandbars, and oceanic whitetips are all carcharhinids. Blue sharks are often seen at or near the surface, and are probably the most frequently observed by sailors. They are cobalt blue above and white below, and can be identified by their slim shape, long, sickle-shaped pectoral fins, and a white eye-ring, which gives them a staring, wide-eyed appearance. The tiger shark, named for its dark stripes on a yellowish-brown ground, is also easy to identify: It has a squared-off snout, and a tail where the upper lobe is much longer than the lower. (This is known as a "heterocercal" tail; those sharks with a quarter-moon tail, where the upper and lower lobes are approximately of equal size, are called "homocercal.") Bull sharks *(Carcharhinus leucas)* frequent inshore waters and sometimes even swim far upstream in freshwater rivers. (At one time, there were believed to be distinct species in various rivers, such as the Ganges shark, the Zambezi shark, and even the Lake Nicaragua shark, but they have all been shown to be our old friend, the bull shark.) Bull sharks are probably the most dangerous of all sharks, since they are the ones most likely to be found in areas where people swim. A series of attacks in South Africa in the 1960s have all been attributed to bull sharks, and the celebrated four killings (and one severe mauling) that took place off the New Jersey coast in the summer of 1916, long attributed to a "rogue" great white, have now been assigned to bull sharks. Duskies and sandbars are often mistaken for bulls, since they are all rather nondescript grayish-brown sharks, reaching a maximum length of about eight feet, but the dusky has a different pattern of denticles (the toothlike structures that comprise the abrasive skin of sharks), and the sandbar *(Carcharhinus plumbeus)* has a much higher, more pointed dorsal fin. The oceanic whitetip, usually found further offshore, can be identified by its rounded dorsal and pectoral fins, which are tipped with white.

The hammerheads, of which there are at least nine distinct species, are probably the most easily recognizable of all sharks. To a greater or lesser extent, they all have elongated, flattened lobes on their heads, on the ends of which are located their eyes. These "ham-

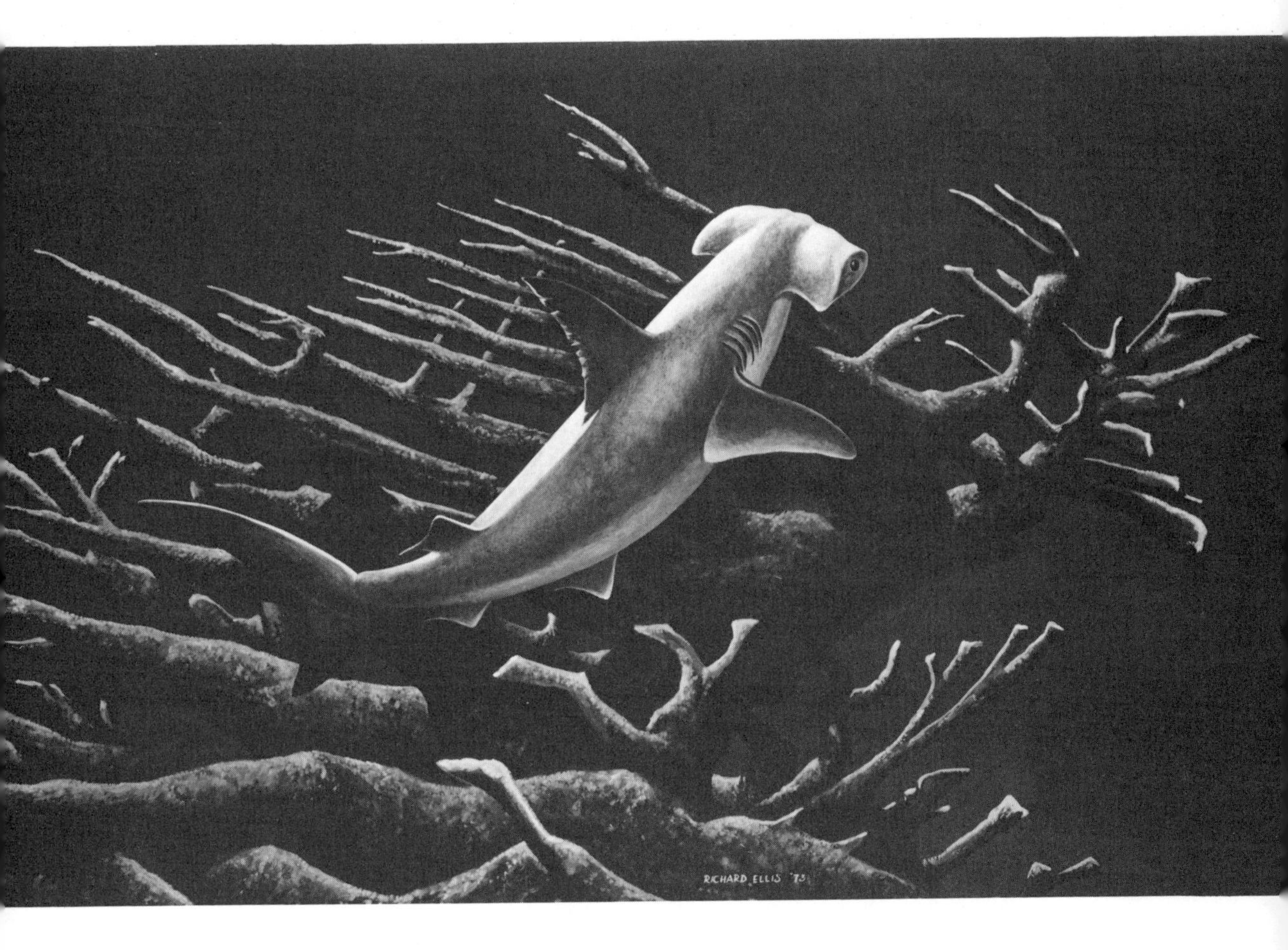

Hammerhead shark RICHARD ELLIS

mers" have been the subject of much speculation, since they are unique; they have no homologues in nature. It has been suggested that the lobes serve as balancing planes for turning or diving, or even to allow the shark to increase its binocular vision by placing the eyes farther apart. The actual explanation (although it is still only a hypothesis) is probably much simpler. All sharks are sensitive to electrical fields, which they can detect through a series of jelly-filled pores on the underside of the snout, called the Ampullae of Lorenzini. Since the hammerhead feeds mostly (but not exclusively) on sting rays that lie buried in the sand, it sweeps its head slowly from side to side as it cruises over the bottom, picking up the electrical impulses from its potential prey. It stands to reason that the widest reception area would be the most efficient, ergo, the hammerheads' hammer.

The great white is one member of the genus known as the Isuridae, or mackerel sharks. Also included are the makos (longfin and shortfin), and the peculiarly named porbeagles. All the mackerel sharks are fast-swimming, streamlined sharks, with sharply pointed snouts and homocercal tails. They all have characteristically flattened keels at the base of their tails, a modification thought to add to their swimming power and speed. (The porbeagle has a secondary keel below the main one, which, while it serves to identify the species, does not perform any known function for the shark.) Makos, which are a rich, dark blue above and snow-white below, reach a maximum length of 12 feet, and can weigh over 1,000 pounds. Like the 8–10-foot porbeagles, they swim with stiff, short strokes of their powerful tails, unlike the sinuous movements of the carcharhinids.

There are dozens more in the fascinating family of sharks; some lie on the bottom, and others glow in the dark. There are sharks known as threshers which have tails as long as their eight-foot-long bodies, and which are believed to herd their prey with sweeping strokes of these incredible tails. There are also angel sharks, goblin sharks, carpet sharks, swell sharks, nurse sharks, monk sharks, bramble sharks, silky sharks, cow sharks, frilled sharks, cat sharks, dogfish, and wobbegongs.

Thresher shark RICHARD ELLIS

Obviously, the primary problem that people have with sharks concerns the attacks. These attacks have never been very numerous, but they are often so grisly that they receive publicity all out of proportion to the actual event. All kinds of things are statistically more dangerous than shark attacks, for example, bee stings, lightning strikes, and of course driving a car. (It has been suggested that it is more dangerous to drive away from a beach where sharks have been reported than it would have been to swim there.) But still the horror stories prevail, and there has been an inordinate number of man-hours devoted to solving the mystery of why some sharks attack some people.

The first full-scale assault on this problem took place during World War II. When aviators and sailors found themselves accidentally dumped into the ocean, they were often the victims of shark attacks. Therefore, the Navy searched for some kind of repellent that would keep the sharks away from the floating people. A "Shark Chaser" was developed, which consisted of a packet containing a black dye that would cloud the water and a copper acetate solution that was thought to inhibit the sharks' feeding responses. Shark Chaser was developed in accordance with the prevailing theories that the sharks were trying to eat the people in the water, an assumption that has now been discarded. (We still do not know why sharks bite us, but if they were interested in eating us, no beach in the world would be safe for swimming.) Various other devices were developed to keep sharks away from people, including bubble screens, plastic bags, and other substances that were believed to repel sharks. None of them worked particularly well, and even when a substance was isolated from a Red Sea flatfish known as the Moses sole which seemed to prevent a shark from closing its mouth, it was found to be too complicated and expensive to manufacture. Divers have even tried chainmail suits, but these are far too cumbersome for the average swimmer.

Sharks are attracted to noise in the water—they are among the most acoustically sensitive animals on earth—and therefore, many of the suggestions that had been made in earlier, simpler times ("thrash around to frighten the sharks away," "shout at them underwater") had exactly the opposite effect: Noise *attracts* sharks to the scene.

Because there is no single "shark," there is no single device that will protect swimmers. What frightens off a reef shark might attract a great white. The obvious solution to the problem is to stay out of the water altogether, but this is not possible for shipwrecked sailors or for those who would enjoy recreational swimming. In 1982, a windsurfer was sailing along in Hawaiian waters, when he fell off his board. He was immediately bitten in several places on the leg by an unidentified species of shark, and, at first, it was assumed that the shark had been cruising along after him, just waiting for him to fall off. A more rational explanation is that the shark happened to be in the immediate vicinity, and when the surfer fell in, the innocent shark attacked as much out of surprise as malice. Attacking sharks may also confuse a swimmer or surfer with their normal prey, or they may be more protective of their territory than we had previously assumed. There may still be reasons for shark attacks that we do not understand.

If you are in the water (uninjured), and a shark appears, by far the best thing to do is to get out of the water if it is at all possible. If you cannot do this, do not thrash around, do not splash or shout, and do not punch the shark in the nose. If you are injured, try not to bleed. It is not true that sharks are driven into a frenzy by the presence of blood in the water; if they were, they would go crazy every time they ate. However, sharks have an acute sense of smell, and they can identify blood in the water from great distances.

Above all, sharks demand respect. They are not cruel and hungry man-eaters, but simply (or perhaps not so simply) predators that usually feed on large prey. Many of them are graceful and powerful, but they are not particularly intelligent. They do, however, perform the business of being a shark with great skill. If you would enter the sharks' element, remember that they have been there for over 300 million years. They were there first, and they belong there. As outsiders, who cannot even see underwater without the aid of artificial devices, we must give the sharks their due. Do not shoot them, harpoon them, or harass them. They often ask no more than to be left alone, to swim as they have always done, in the serenity of their own environment.

SUGGESTED READING

The Lady and the Sharks, by Eugenie Clarke (New York: Harper & Rowe, 1969).

The Book of Sharks, by Richard Ellis (New York: Grosset & Dunlap, 1976).

The Natural History of Sharks, by T.H. Lineaweaver and Richard H. Backus (New York: Lippincott, 1973).

Richard Ellis is considered the country's foremost painter of marine natural history subjects. He is author/illustrator of the highly acclaimed Book of Sharks *(1976),* The Book of Whales *(1980), and* Dolphins and Porpoises *(1982). He has dived with sharks and whales all over the world, from Newfoundland to Patagonia, and has served since 1980 as a member of the United States delegation to the International Whaling Commission. He is currently working on the third volume of his ambitious work on the marine mammals of the world, which will include seals, sea lions, walruses, manatees, and polar bears.*

SEABIRDS

Dr. Donald Bruning

Gannet

Seabirds and man have interacted ever since man ventured out onto the sea. Primitive man marveled at seabirds and used them to his benefit. The earliest sailors recognized that birds could give definite and often vital information about distance and direction of land. They also looked to birds for predictions of storms and other weather changes. Early fishermen learned that groups of seabirds feeding meant a school of fish near the surface.

Even today, fishermen carefully watch seabirds to locate fish. Watching a large group of seabirds wheeling in flight and then diving into a school of fish can be one of nature's most spectacular sights. Frequently, several species of birds are involved, as well as porpoises. These feeding frenzies are a sight one never forgets.

Cormorant

Birds like gannets, boobies, and cormorants dive deep beneath the surface for fish, while gulls and terns catch fish at or near the surface. Even the solitary diving of pelicans and terns can be a beautiful sight, but the grace and speed of diving gannets is unmatched by other birds of the Atlantic. Today, with the transition to modern navigational equipment, most sailors no longer use birds to guide them. However, with just a little knowledge of birds, many modern-day sailors can derive

Herring gull

great enjoyment and—in an emergency when other navigational equipment fails—this knowledge may be instrumental in saving their lives.

Most people immediately think of gulls when they hear the term "sea bird." Some gulls are definitely seabirds, but others spend their entire lives far from the sea. "Seabird" generally refers to birds one would see from a ship at sea. These birds come in a variety of shapes and sizes, ranging from pelicans to petrels and albatross to dovekies.

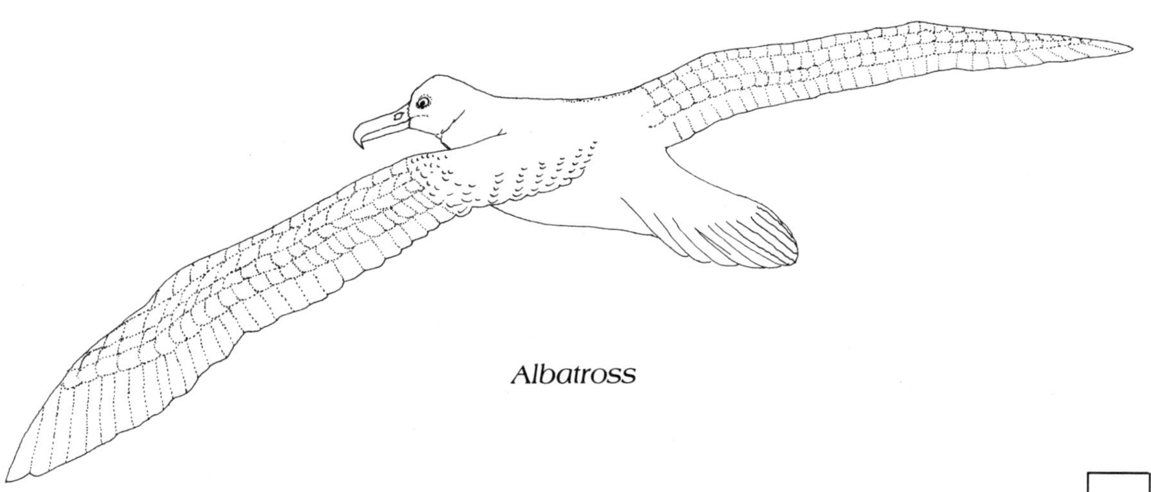

Albatross

Albatross are the largest of seabirds with the largest wingspan of modern-day birds, ranging up to twelve feet from wingtip to wingtip. They are found in the Pacific and South Atlantic but rarely are seen in the North Atlantic.

Petrels, like albatross, are truly seabirds that spend almost their entire lives at sea. They return to land once a year for nesting. The North Atlantic is home for a number of members of the petrel family, including several species of shearwaters, storm petrel, fulmar, and true petrels. This entire group of birds is frequently called tube-nosed because of the obvious external nares opening on top of the bill. These external nares are part of a special salt-removal system which allows these

Petrel

birds to exist drinking only saltwater. Without this special adaptation, these birds would not be able to survive as true seabirds.

The pelican and its relatives, the gannets and cormorants, are among the most spectaculr of Atlantic seabirds. The brown pelican is basically a coastal bird that is rarely seen far out at sea or inland. These pelicans are marvels to watch as they seem to fly in formation and make plummeting yet graceful dives to catch fish. In mid-flight, they fold their wings, extend their long necks and bills, and plunge into the water. Their bills and pouches serve as a huge net for catching fish. After a dive they bob to the surface, expelling all the water from their bills and pouches, then proceed to swallow their catch. These birds are well adapted for this type of fishing, with hollow bones and an elaborate air sac system that keeps them very light and very buoyant while also serving as a shock-absorber system during their dives.

The brown pelican very nearly disappeared in the 1960s due to the pesticide DDT which interfered with their calcium metabolism, resulting in very thin-shelled eggs. The plight of these beautiful birds helped provide impetus for the ban on use of DDT in the United States. Fortunately, most of the adult pelicans survived the DDT and slowly are beginning to reproduce along the southeast and gulf coast of North America.

The gannets and cormorants can be seen far out at sea in northern waters, although they are normally found along the coast—particularly around offshore islands. Large gannet colonies, like those on Bonaventure Island, Nova Scotia, are spectacular sights.

Tropicbirds are other close relatives of the pelican. Because they are found in tropical and semitropical seas, the likelihood of seeing tropicbirds increases as one travels south in the Atlantic. Tropicbirds are quite common in parts of the Caribbean and generally are found in warm coastal waters, but frequently can be seen just out of sight of land.

Gulls and their close relatives are mainly coastal birds—most species are seldom seen out of sight of land. The kittiwake of the North Atlantic is an exception and can be seen far out at sea except when breeding along the coast during the summer months. Herring gulls will occasionally follow the kittiwakes well out to sea, and flocks of lesser black-backed gulls can be seen in the open ocean during spring and fall migration. Four Skua species, gulls that snatch food from other seabirds, occur in the North Atlantic during the spring and summer. However, during the winter only three species can be seen as the parasitic Jaeger moves south to the tropics or southern hemisphere.

Most shorebirds, true to their name, venture to sea only rarely or in a bad storm. Phalaropes, wading birds, are an exception, as they fre-

Pelican

Tern

quently fly over expanses of open ocean during spring and fall migrations. In recent years, a great deal of research has been conducted on bird migration and navigation. To date, only a few of the secrets have been discovered. While most birds use visual landmarks, many species that migrate at night, or far out at sea, have none to use. The sun, the stars, and the magnetic forces of the earth have all been implicated as navigational aids for birds; unfortunately no single method seems to work for all birds. An observant sailor knows that he can generally follow any of the coastal birds to land, but following a true seabird may not lead him to any shore.

The auk family is well represented in the Atlantic. Several species of puffin, murres, guillemot, razorbill, and dovekie are a few examples. Most auks nest on islands or rugged coasts in the North Atlantic, but many can be seen at sea, some distance from land, while feeding.

Terns are well represented in the North Atlantic during the summer but retreat to warmer waters during the winter. Terns also can be seen only occasionally in the open ocean.

Watching birds for sheer pleasure is a sport enjoyed by more and more people of all ages. All that is necessary for this enjoyment is one's eyes and ears. A good pair of binoculars and a field guide are a great help. The more avid enthusiast needs a spotting scope, checklist, and more field guides.

As people learn and benefit from birds, so birds use people. All one needs to verify this is to watch fishing boats as they are returning to port after a successful day on the water. When fish are being cleaned, the flock of gulls that follows such boats can be spectacular. Ships, several hundred miles out to sea, also attract the petrels and shearwaters that frequently sail for hours over the wake of a ship, waiting for anything dumped into the water or stirred to the surface.

Birds can give hours of pleasure to anyone who takes the time to watch them soaring effortlessly and gracefully over the waves.

Puffin

SUGGESTED READING

Birds of the Ocean, by W. B. Alexander (New York: G. P. Putnam's Sons, 1954).

Birds of North America, by C. Robbins, et al. (New York: Golden Press, 1966).

A Field Guide to the Birds, by Roger Tory Peterson (New York: Houghton Mifflin Company).

Donald Bruning was born in Boulder, Colorado, and grew up on a nearby farm. He attended UCLA and Colorado University, where he received an M.A. and a Ph.D.

Mr. Bruning began his zoo career in 1967 as ornithologist at the Bronx Zoo. He became assistant curator in 1969, associate in 1973, and curator in 1975. He has been supervisor of New York Zoological Society travel programs since 1977, scientific adviser to the World Pheasant Association since 1979, NYZS representative on the council of the Bahamas National Trust since 1979, and adjunct professor at Fordham University since 1974. He also served as a research associate for the Zoological Society's center for field biology and conservation from 1973 to 1979.

Mr. Bruning joined the American Association of Zoological Parks and Aquariums in 1968 and is currently serving as chairman of the legislative committee. He served on the honors and awards, legislative, and program committees of the American Association of Zoological Parks and Aquariums. He served as chairman of the program committee in 1976. Mr. Bruning is a member of the American Ornithologists' Union and over a dozen other professional and environmental organizations. He has traveled extensively as a tour leader for NYZS and has conducted research in South America and Asia. He has authored over fifty technical and popular articles and presented papers at over twenty professional and hobbiest meetings.

The FACTS of FISH

Jerry Kenney

The ocean is the most fertile area on the earth. Life had its beginning there, and while most forms have evolved over the ages, some have not changed at all.

Our western Atlantic Ocean harbors something like 3,000 species of fish. So when we launch into it in our little boats, we should understand that we're trespassing on some very valuable property.

The average recreational boater doesn't have any idea what's going on under him when he's out there. The finny residents lead pretty secretive lives. Occasionally though, they put on a show, and when they do, it's usually spectacular. And if we're out there and we keep our eyes open, we're going to see it.

But it helps to know what we're looking for. I spent three years on ships in the Navy, and while I saw a lot of water in a couple of oceans, I didn't see much in the way of marine life merely because I wasn't looking for it. Later, as a civilian sailor, I soon became aware the ocean was more than just deep water.

Certain regions along the coast harbor different species of fish. We'll see twenty off the New York Bight that we won't see off Cape Hatteras. And Florida waters have hundreds of species, particularly in the Keys, which are surrounded by live coral reefs and are themselves accumulations of dead coral.

In most cases, fish remain relatively local. Few travel more than a couple hundred miles and often much less from their home grounds, unless they are of a pelagic species.

What keeps these local fish from mingling with those of another region? Well, distance is certainly a barrier, but mostly it is water temperature.

In the case of bluefish, they are caught off Florida as well as off New York and Maine. But they aren't the same fish. The Florida blues travel no farther north than Cape Hatteras or the Carolinas. And this is where the blues we have up in the Northeast come in from open ocean and head north. They never go farther south than Hatteras before again heading for open sea in late fall.

Snook, king and Spanish mackerel travel great distances in the Gulf of Mexico, around the southern tip of Florida and up the coastline. Channel bass and drumfish will migrate up from Florida to the Outer Banks and Delaware Bay as the water temperature increases.

Atlantic or Boston mackerel come inshore off Cape Hatteras from mid-ocean wanderings and start a northward migration in April. They, too, follow the rising water temperature all the way up past Montauk,

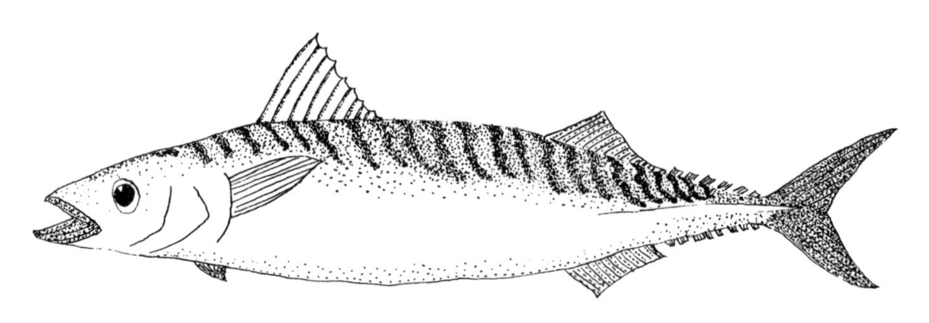

Mackerel

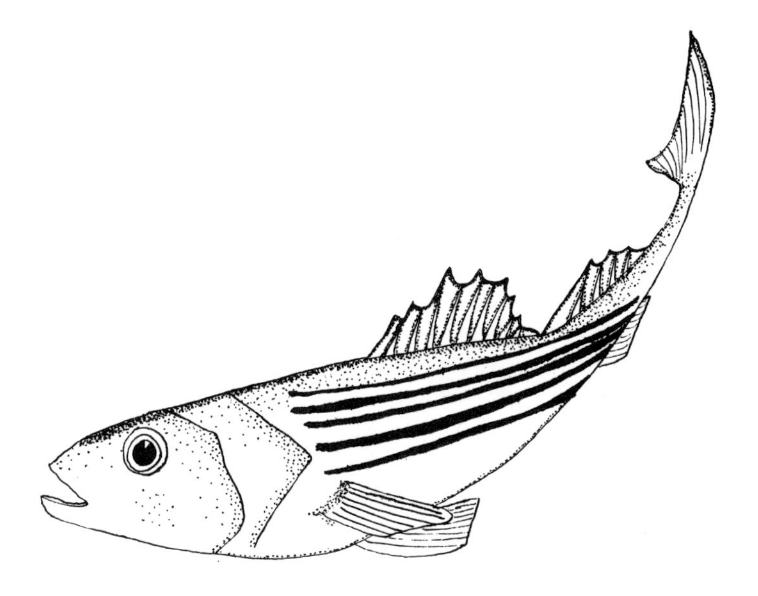

Striped bass

New York, to New England. And there are more than is humanly possible to count—many millions. As they arrive in the New York-New Jersey Bight, the schools often stretch from Manasquan Ridge to Ambrose Light, a distance of thirty miles.

During this period fishermen on commercial and recreational party, charter, and private boats catch these fish, which average one to three pounds, by the millions. And there are still more.

And as the mackerel move along in this massive body, bluefish are right on their tails. They also come by the millions in loose schools covering vast stretches of coastal areas.

Shorter migrations are made by species such as striped bass and weakfish. Stripers are the most prized inshore gamefish along the north Atlantic coast, even though they seem to be suffering a population setback from pollution and overfishing. The Chesapeake and Delaware bays are major breeding areas for bass, along with a section of the Hudson River about twenty miles north of the George Washington Bridge.

Sharks and tuna are believed to have the longest migrations. Tuna, for example, have traveled as far as the coast of Norway and down into the Mediterranean before circling back. Then they will visit the Bahamas, the Gulf Stream between there and Florida, and then proceed northward to the waters off eastern Canada, New England, and Montauk, New York. Generally, their final appearance of the season is in the Mud Hole area. This is a deep gorge off the coast of New Jersey that is a continuation of the Hudson River. It's only about fifteen miles from shore, but about sixty miles further out at the edge of the continental shelf, it becomes the Hudson Canyon.

Observing surface patterns on the ocean is probably the most effective method of spotting fish. Diving gulls and seabirds are also a

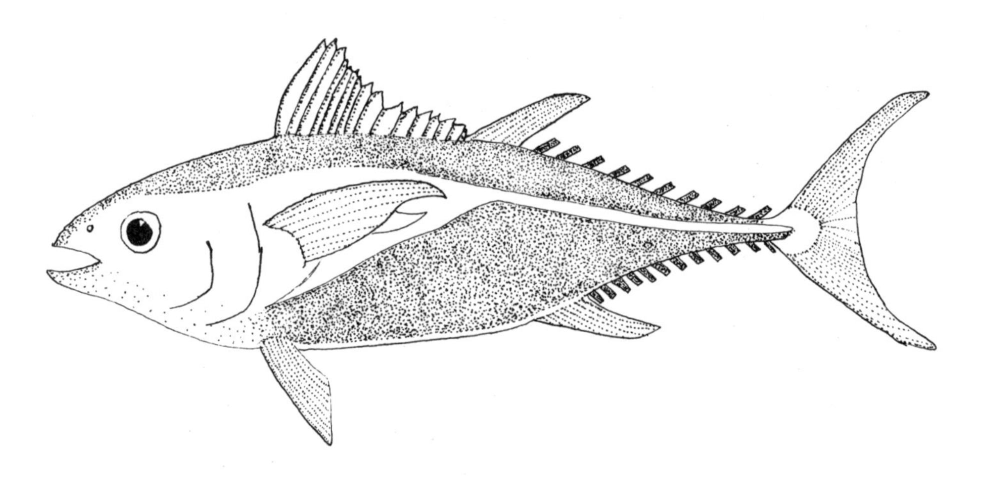

Tuna

dead giveaway. If they are hovering over a particular area or diving frantically at the surface, it's a cinch they're feeding, probably on small fish that they can pick up and swallow. But generally the reason those bait fish are close to the surface in the first place is that they are being pushed up and attacked from below by larger predators.

Some predators will be easily seen as they cruise the surface. Shark, marlin, sailfish, and broadbills often expose their dorsal and tail fins while near the surface.

Bottom configuration and depths determine the texture of surface water. A swift-moving tidal current flowing through a deep gorge will often create telltale surface patterns different from the surrounding area. Currents running up against shoals or bottom obstructions cause waves, ripples, and swirls, and wherever there is commotion of this type, fish are generally in the vicinity.

I suppose the easiest way to see or locate fish is by watching for clusters of fishing boats. These days, with the sophisticated fish-finding boat equipment, it's easy to spot fish on a screen, and there's nothing wrong with letting someone else find the fish for you.

Further offshore, in the blue water, there may be no obvious signs of life, but a simple weed line or floating debris will often attract fish, such as billfish, dolphin, and tuna.

Basically, any area of surface water that is different from surrounding areas indicates something's going on down below. It could be simply a current. This churns up the bottom, which in turn washes up crustaceans and plankton, which are followed by bigger fish in the ongoing food chain.

It may take a while for newcomers to identify the fish they see. But there are certain species that are indigenous to certain areas, and it's likely you will come in contact with these fish regularly. Identification comes with experience. If you see one striped bass or bluefish or dolphin, you will always remember them. Southern bottom fishermen will have a tougher time identifying the various strains of groupers and snappers and cunners. A king mackerel will stick in one's memory as will a cobia, because these are exciting fish to see and catch. The hulking jewfish, common at 300 pounds but weighing up to nearly 700 pounds, are not likely to slip your memory either. Barracuda, like sharks, are among the most photographed species, and you won't forget them.

Swordfish, blue and white marlin, and sailfish, which have bills or swords protruding from their heads, use their weapons to thrash into schools of fish. They then swim leisurely around inhaling the dead and wounded.

A frightening critter if there ever was one is the sawfish, found only in tropical waters, rivers as well as oceans. It has some of the characteristics of swordfish, shark, and rays. The body is sharklike, but flattened like a ray, and the sword it brandishes is studded with spikes every two or three inches.

One monstrous-looking fish that inshore boatmen are likely to meet sooner or later is the angler fish, sometimes called the goose or monk fish. This guy is an ugly blob of brown slimy skin weighing anywhere up to forty pounds, but virtually all mouth. Its enormous maw is armed with rows of needlelike catcher teeth, and it also sports its own fishing

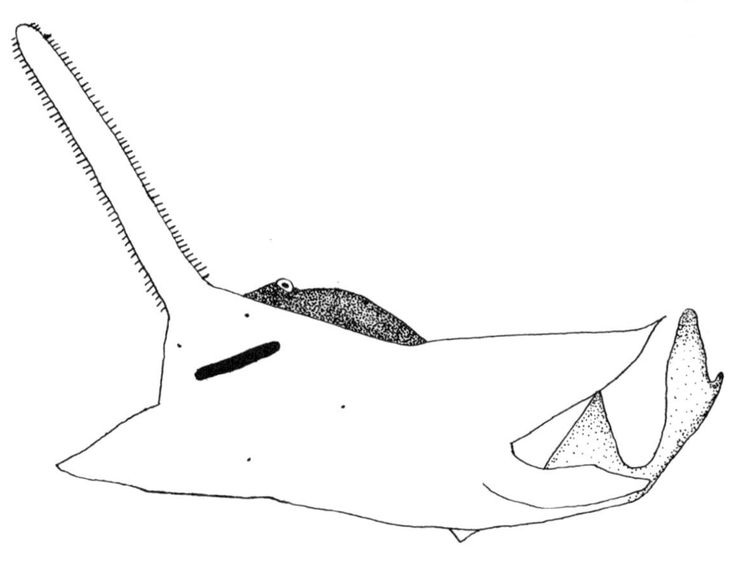

Saw fish

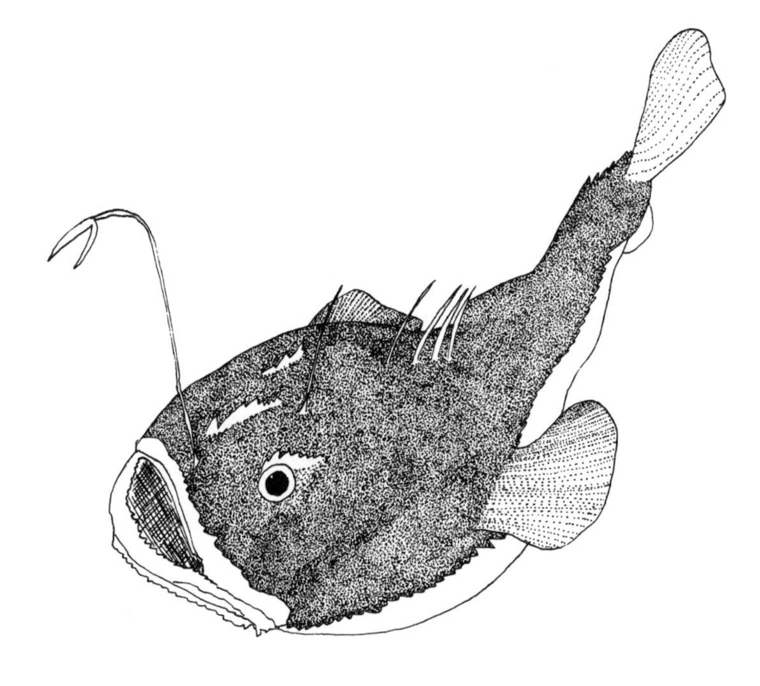

Angler

pole and bait. The angler fish lies flat in the mud, which it looks to be part of, with its mouth wide open. Its frontmost dorsal spine extends over the mouth with a blob of flesh dangling from its tip like a fishing pole and lure. The lure attracts fish, and when they come to investigate, those big jaws clamp shut. Birds have also been found inside angler fish.

The mola mola, or giant ocean sunfish, is frequently seen off our shores in summer, often within a mile of crowded beaches. It can be over 1,000 pounds, but it looks to be only half of a giant fish. The tail seems to have been cut off, leaving this big ugly head swimming around by itself. When on the surface, it is easy to spot from its dorsal fin, which flops from side to side while swimming. As ugly as it is, though, it is harmless and worthless. There is no reason for anyone to harm or attempt to subdue one of these gentle giants.

The blowfish, or puffer, has to be one of our most unusual fish. Not a beauty to start with, it has spines on its skin that in some areas are more like porcupine quills. It has the ability to puff itself up, a defensive technique to make itself look larger and frighten off predators. When caught by fishermen, all it needs is a couple of scratches on the belly and it puffs up like a balloon.

Those who do their boating on tropical waters will frequently see flying fish. These fish really don't fly. They launch themselves out of the water with powerful and rapid thrusts of the long lower caudal fin. The

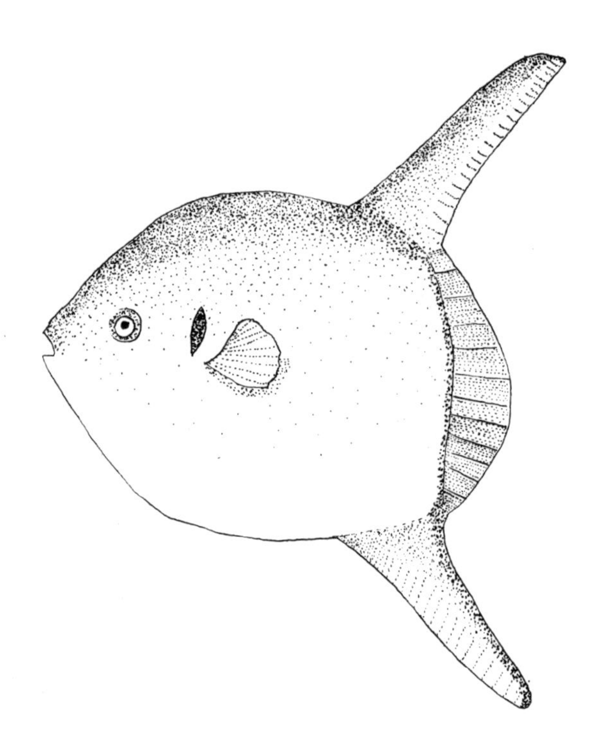

Mola Mola or ocean sunfish

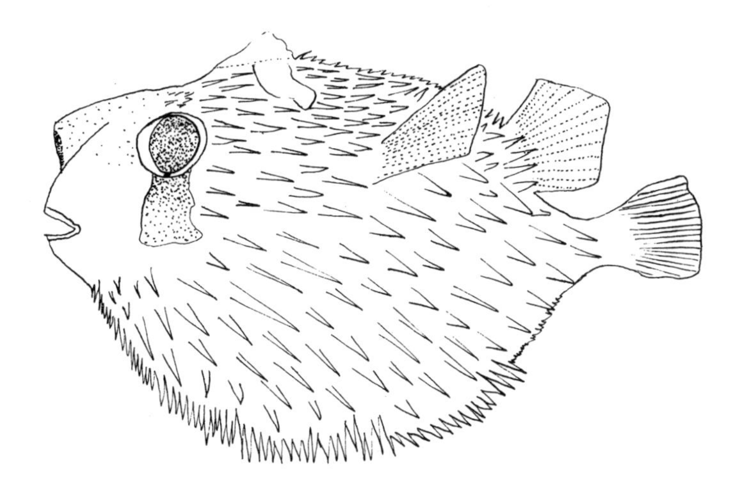

Puffer

tail flutters along the water's surface like a tiny motor as the fish glides for great distances with wings that are actually overdeveloped pelvic or pectoral fins.

Flatfish or the flounder types, fluke, flounder, halibut, and plaice, are so prevalent off our shores that we forget what oddities they actually are. They certainly look strange with both eyes on the top of their flat bodies, even though the mouth notches both top and bottom. But when these fish hatch they look like any ordinary upright swimming fish, with one eye on each side of the head. When they start to mature, though, the fish begin to swim on their side. One side eventually becomes the bottom, the other the top, and the eyes migrate around to the top of the head while the mouth remains where it is.

But even among the flatfish, there are differences. Winter flounder off the northeast coast, for instance, are left-handed, while the summer flounder or fluke are right-handed. That is, the mouth is on the left side of the flounder and on the right side of the fluke. And these flatfish, like many other species, have the ability to change color like a chameleon.

Lying on a dark bottom, a flounder will change to match that color, providing a camouflage that hides him from his enemies.

A dolphin in the water is a gorgeous iridescent combination of blue, green, yellow, gold, and pink, but this rainbow hue remains for only a few minutes after the fish is removed from the water.

Coral fish are the most colorful in the world, and many have the ability to change colors. Even striped bass, blues, and, in particular, weakfish come out of the water in their full array of colors. But one can see the colors fade as life ebbs from the body.

The majority of fish found in the ocean go through a typical life cycle, starting as an egg laid by the female, being fertilized by the sperm of males. Depending on the species, the number of eggs, whether buried in nests in the bottom or spewed freely in open water, can vary from many millions to a few thousand. And it seems the fish that produce the most eggs pay the least attention to the offspring. It's also not unusual for the adult fish to feed on the fertilized eggs, even their own. Other species that produce fewer eggs seem to take very special care of their offspring.

Some fish that are found in the ocean actually only take up temporary residence there. Anadromous fish such as salmon, striped bass, and shad are born in freshwater far upstream from the salty oceans. They spend a little time in the freshwater environment but eventually descend to the ocean. Then they make annual or occasional spawning runs back to the streams in which they were born and, in the case of the salmon and shad, frequently die after successfully spawning once or twice.

Catadromous fish such as eels are born in the ocean and live in freshwater. They spawn southwest of Bermuda. There the eggs hatch, develop into elver which eventually swim towards shore and make their way hundreds of miles back to the freshwater rivers and streams their parents lived in, until they mature. Then they once again repeat the process and die, once spawning is completed.

Fish are mature when ready to take part in the reproduction process. This doesn't mean the fish is fully grown, because fish never reach that stage. Unlike landlubbers, fish continue to grow as long as they live, not quite as rapidly in advanced years as in youth, but the process never stops.

And we've wondered how big some fish get. The largest fish of all is the whale shark, which can attain sixty or seventy feet and weigh up to ten tons. Ironically, this mammoth creature is harmless, and it gets to be this size by eating the smallest sea creatures in the ocean, plankton. Basking sharks, also harmless and huge, feed on the same small animals.

Great white sharks have measured 20 feet and weigh over 5,000 pounds, and bluefin tuna, up to 1,500. Of the inshore species that we are most familiar with, the record books show the biggest caught on rod and reel include blues, 31 pounds; striped bass, 78 pounds; cod, 81 pounds; jewfish, 680 pounds; swordfish, 1,182 pounds; tautog or blackfish, 21 pounds; snook, 53 pounds; and tarpon, 283 pounds.

Fish migrations have been known to scientists for centuries, but the exact routes the fish took was something we couldn't pin down until fish-tagging was developed. Basically, this consists of a dart that can be "stuck" into a fish by any angler with a ten-foot pole. Stuck in a tuna's or shark's back, the dart is harmless, but it bears a durable plastic tag requesting the party that recaptures the fish to return the tag, with date and location of where the fish was caught, along with dimensions and weight if possible. Several local, state, and federal agencies monitor these programs; some even give rewards for returned tags.

In addition to tracing migration routes, this tagging process has also provided evidence that certain species may need better management. Through tagging records, bluefin tuna, for instance, were found to be heavily exploited. With this data, an international management program to help rejuvenate the tuna stock was prepared and is now in effect.

Tagging has also aided in research of life spans of fish and growth rates. The National Marine Fisheries Service in Miami has an active tagging program for billfish and tuna, and conducts a shark-tagging program out of its Narragansett, Rhode Island, base. Any sportfisherman can participate in these programs.

An international tagging program for almost eighty species, including shark, blues, striped bass, fluke, tuna, snook, tarpon, and red snapper is conducted by the American Littoral Society located in Highlands, New Jersey. The only requisite to joining the Littoral program is membership in the society, which is fifteen dollars a year plus the minimal cost of the tags.

A more sophisticated form of sonic tagging has been developed and is being used by the Woods Hole Oceanographic Institute in Massachusetts. This requires transmitters that are stuck into shark, tuna, and billfish, and that their movements be monitored for up to a week.

The most dramatic use of sonic tags recently occurred off Long Island, when researchers managed to "stick" a great white shark with a beeping dart and then track its movements over several hundred miles. A running live account of its wanderings was recorded in the media for three days, when batteries gave out. The shark was last heard from about 150 miles southeast of Montauk, Long Island, while on an easterly course.

SUGGESTED READING

Fishwatcher's Guide to West Atlantic Coral Reefs, by Charles C.G. Chaplin (Valley Forge, Pa.: Harrowood Books, 1975).

Guide to Marine Fishes, by Alfred Perlmutter (New York: NYU Press, 1961).

Fishes of the World: Illustrated Dictionary, by Alwynne Wheeler (New York: Macmillan, 1975).

As outdoors editor on the New York Daily News, *the country's largest-circulation newspaper, Jerry Kenney touches all bases. In the winter it's skiing; in the fall, hunting; the spring, camping. But fishing overlaps them all.*

It is mostly a warm-weather activity for him, but he follows the warm weather wherever it may be. His bailiwick is the Northeast—New Jersey, New York, and Connecticut waters—but he has "wet his line" in salt- and freshwater from Maine to Florida, the Bahamas and the Virgin Islands, Central and South America, Canada, England, Ireland, Scotland, and South Africa.

His philosophy regarding his favorite "beat" is: "Fishing gives me peace of mind—and fillet of sole."

The CASE of the DISAPPEARING LOBSTER

John T. Hughes

When you become acquainted with the lobster's life history, it seems amazing that so many of them make it to our dinner tables at all. The Maine lobster, the New England lobster, the American lobster (Homarus americanus) comprise the most valuable marine species landed on the eastern coast of the United States. The species represents about $75,000,000 to lobstermen from Maine south to the Carolinas for the 30,000,000 pounds of lobsters landed. Maine lands the lion's share, Massachusetts the next largest, with smaller shares occurring as the latitude decreases. Canadian fishermen catch more lobsters than all United States fishermen combined—about 60,000,000 pounds.

Lobster populations are declining. The extent of the decline of the lobster population is indicated by the catch per pot (or unit of effort) at the various dates: for example, in 1889 there were 225 pounds of lobsters caught per trap per year; in 1940 there 34 pounds caught per pot; and in 1970 only 17 pounds of lobster were caught per pot. Clearly, in the past twenty years the total lobster catch has been decreasing. More and more lobstermen are fishing with much more gear to catch fewer lobsters. In Massachusetts alone, 4,300 lobster licenses were issued in 1967, and 130,700 pots were fished to catch 3,102,970 pounds of lobsters. However, six years later, in 1973, 9,048 licenses were issued, and 169,749 pots were fished to catch 3,680,554 pounds of lobsters.

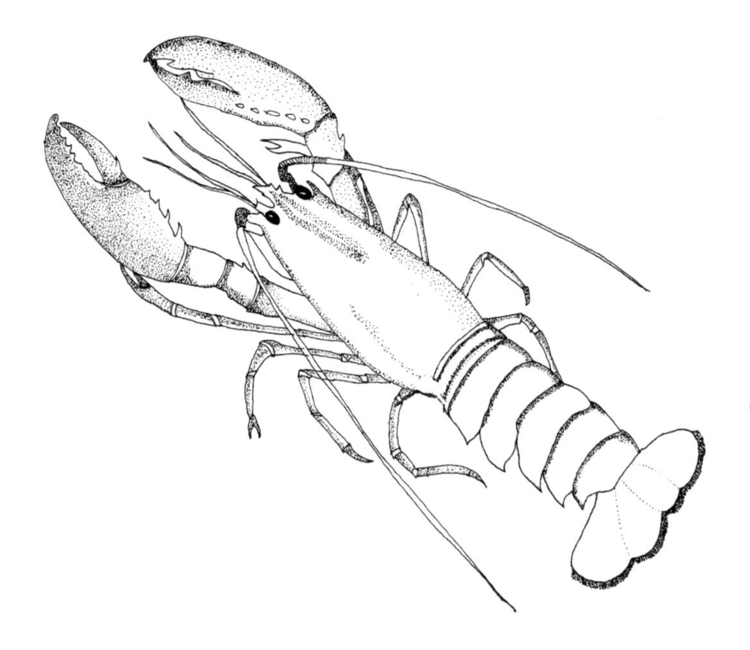

Lobster

Think of it—with more than a 100 percent increase in the number of licenses issued, 30 percent more pots were fished, but a mere 12 percent increase occurred in the landings.

Why has this happened? There are a number of reasons for the decline of lobsters. Lobsters are extremely sensitive to water quality. In addition to overfishing and foreign competition, fluctuating ocean temperatures and pollution play a role. Not only do minimal amounts of oil or kerosene cause them to stop eating, but oil and pesticides on the surface of the ocean can kill the tiny lobster larvae.

Let's take a look at some lobster biology. A one-pound lobster is about seven years old—the age at which lobsters reach sexual maturity. One way to distinguish a male from a female lobster is to compare the width of their tails—the female's tail is always wider. Lobsters mate as mammals do—that is, by physical contact in a head-to-head position with the male uppermost. The only time lobsters can mate is within forty-eight hours after the female molts, or sheds her shell. At time of molting, the mature female extrudes a pheromone that attracts the male. After a "courtship" procedure that may last for as long as thirty minutes, sperm is transferred into the seminal receptacle of the female. Here it is stored for nine months while the eggs prepare for fertilization. At that time, the female extrudes up to 60,000 eggs and cements them to hairs of her swimmerets, under the tail. As the eggs are extruded, the stored sperm is released to fertilize them. Then the eggs remain glued to her swimmerets for a second period of nine months, while cell division and embryo development take place.

Early in the summer, some eighteen months after mating, the lobster larvae hatch from under the female's tail and swim away. At this point, they look more like shrimp or mosquito larvae than lobsters. After four moltings, in a period of about three weeks, the lobster fry look like the adult. However, during the first three stages (molts), the larvae live in a free-swimming state and are unable to sink to the bottom. Bunching up near the surface, they are easy prey to birds, fish, and to each other, for the lobster is cannibalistic. In addition, while living near the ocean's surface, the lobster fry are affected by oily slicks, pollution, and by tides and currents that carry many of them out to sea to perish. Most estimates say that fewer than one-tenth of 1 percent of the larvae reach the fourth stage and become bottom crawlers, after about three weeks. These lucky few seek dark places, lead a nocturnal life, and acquire the defensive instincts of the adult lobster. However, throughout life, even when they reach forty pounds in weight, they have enemies. Not only does man fish for lobsters, but codfish, dogfish, and others swallow lobsters whole.

In order to be taken legally by fishermen, lobsters have to be of a certain minimum size. Any lobster under this minimum size must be returned to the sea at once. This minimum size varies in each state. Instead of the total length of the lobster, the carapace shell—the large back shell that abuts the eyes and tail—is measured. (If total length were used, the fisherman could stretch the body to increase its length.) A sublegal lobster in Massachusetts, for example, is a lobster measuring less than 3³⁄₁₆ inches in carapace length, measured from the rear of the eye socket along a line parallel to the center line of the body shell to the rear end of the body shell. A lobster of this size is just under one pound in weight. About 90 percent of all lobsters caught in the inshore lobster-pot fishery have just reached this minimum legal size. In other words, these lobsters were sublegal before their last molt, for at each molt, lobsters can increase 20 percent in length and up to 50 percent in weight. Of this 90 percent, less than 10 percent have reached sexual maturity or bred even once! Most lobsters taken, therefore, have not had a chance to produce young.

Some unscrupulous fishermen and divers also illegally take lobsters with eggs under their tail, removing the eggs with stiff brushes or with water pressure from a hose. There are laws against this, and methods have been developed by marine biologists to detect lobsters that have had their eggs removed by force. One lobsterman was fined $10,750 and had his vessel seized!

What is now being done to protect the lobster? At present, each state with a lobster industry, from Maine to South Carolina, has its own rules and regulations; law-enforcement efforts also vary with each state. There are no federal rules or regulations dealing with lobster fishery. However, a federal/state effort is under way to standardize lobster management. both federal and state governments realize that the resource is endangered and that steps must be taken while there is still time for them to be effective.

Among the recommendations by the Federal/State Committee on standardizing lobster management are the following: control of total fishing effort on all lobster stocks; a system of reciprocal law enforcement; establishment of a uniform legal minimum size of 3½ inches, the size at which lobsters attain sexual maturity; protection of egg-bearing female lobsters; the use of lobster pots with escape vents to discourage cannibalism and the catching of undersize lobsters; the establishment of standardized licenses and a system of record-keeping to facilitate better reporting of the total catch.

There is yet another method of increasing the lobster population. Many years ago, Canada, Maine, Massachusetts, Connecticut, Rhode

Island, and the federal government had lobster hatcheries to hatch and rear lobster through the perilous free-swimming stages at which time 99.9 percent of lobster mortality takes place. Today, only Massachusetts has a viable lobster-rearing program. In 1939, the Massachusetts state legislature passed a bill to establish a lobster hatchery on Martha's Vineyard. Its purpose was to hatch lobster eggs and rear the newly hatched fry to the bottom-crawling stage, at which time they would be liberated into coastal waters. Since 1949, when the hatchery was first in production, Massachusetts has liberated millions of lobsters into the fishery. It has been difficult to determine if this release by the Department of Natural Resources has had an effect on the harvest, for, since lobsters have to shed their entire shell to grow, there is no way of marking or tagging these baby lobsters so that they can be identified when captured.

A lobster's natural color is a blend of its normal pigments, red, yellow, and blue, which results in a dark greenish-brown coloration, but in the holding tanks for the laboratory, there are dozens of freaks of nature. There are bright blue lobsters; scarlet, calico-colored, and spotted specimens; lobsters with one red and one blue claw; and myriad combinations of red, white, and blue.

Lobsters can be bred to produce strikingly different colorations in the offspring. A bright red lobster is the result of mating a normal-colored male and a red female. From that union, 25 percent of the young are the usual natural color, 25 percent are albino, and 50 percent are red. The shell of an albino lobster completely lacks any coloration.

These lobsters will be color-coded for life; their distinctive markings will remain no matter how many times they shed their shells. Lobstermen are being asked to notify the lab, (Lobster Research Station, Box 894, Vineyard Haven MA 02568,) of any unusual specimens they find in their traps. In this way, we can trace the lobsters' migratory habits, learn what kind of sea bottom they prefer, and how far they wander from their original spawning grounds. Scientists must continue to look for clues as to why the lobster population is declining.

Tagging and tracking lobsters is an important part of the whole lobster-fishery scene. Armed with the information they receive, scientists can predict such things as future catches and why lobsters are plentiful in certain areas one year and scarce the next. With enough information, they can draw maps and ascertain the entire picture of the lobster population and its movement.

SUGGESTED READING

About Lobsters, by T. M. Prudden (Freeport, Maine: Wheelwright, Bond, 1973).

Natural History of the American Lobster, by Francis H. Herrick (New York: Arno Press, 1978).

John T. Hughes has been the chief of the Lobster Research Station in Vineyard Haven, Massachusetts, since 1975. Before receiving degrees from the University of Massachusetts and Cornell University, Mr. Hughes commanded a minesweeper in World War II. In 1973 he was named "State Scientist of the Year" in Massachusetts. He has received the International Award for "Outstanding Service in the Aquatic Sciences," the citation and scroll from the governor of Massachusetts "In recognition of his pioneering research on Lobsters," and has an international honorary life membership in the World mariculture Society.

Mr. Hughes has served as citizen ambassador to the Peoples' Republic of China for bilateral exchange with Chinese specialists in aquaculture. He has also consulted to programs in Italy, Spain, France, Ireland, St. Croix, Bermuda, Puerto Rico, Grand Turk, Tahiti, Nassau, Walker's Cay, Turkey, Japan, Australia, and Brazil. He can be found on the roster of nearly every major advisory board in the field of aquaculture, and has been published over twenty-five times since 1957.

INVERTEBRATES

Dr. Stephen Spotte

To a biologist, all animals without backbones are *invertebrates.* Those that inhabit the oceans are *marine invertebrates,* and the best place for a boater to observe them is right at the dock. Typical terrestrial invertebrates are insects, spiders, centipedes, and scorpions. Marine invertebrates most likely to be seen by peering down into the water include sponges, sea anemones, sea stars, sea squirts, and many different types of mollusks and crustaceans. Marine invertebrates range in size from single-celled animals so tiny they are invisible without the aid of a microscope to giant squids that attain lengths of twenty meters (sixty feet).

Like all living organisms, marine invertebrates exist in *biomes,* broad communities of animals and plants that characterize major habitats. Terrestrial biomes include the vast grassland or veldt of East Africa, tundra like the Brooks Range of Alaska, and different tropical rainforests. More restricted biomes also exist, and these are easily recognized by the boater. On the Atlantic coast the three typical biomes of marine invertebrates are *(1)* pelagic, *(2)* balanoid-thallophyte, and *(3)* pelecypodannelid. The pelagic biome consists of animals and *plankton* that drift with the currents. Planktonic animals and plants likely to be encountered by mariners are covered in the chapter on plankton by George D. Ruggieri.

Balanoid-thallophyte translates literally to barnacle-seaweed. On the Atlantic coast this biome is most strongly represented in New England, where boulders and headlands provide attachment sites. The most conspicuous feature of the balanoid-thallophyte biome is *intertidal zonation,* which causes rocks and pilings to resemble the sides of a bathtub with different-colored rings. This effect is apparent only at low tide. Close examination shows how the rings form. As might be expected, barnacles and seaweeds account for a sizable number of rings. The topmost section of the balanoid-thallophyte zone, which is well above the high-tide mark, is always dry except for occasional seaspray. This zone is conspicuous for its colorful *lichens* that form encrustations. Lichens consist of symbiotic colonies of fungi and algae, and therefore are classed as plants. The black line, constituting the dirtiest ring on the bathtub, is comprised of *cyanobacteria* (also called blue-green algae), which are colonies of microscopic plants. The first animals are the *periwinkles,* thumbnail-sized snails that feed on encrusting algae. Next come green seaweeds, then *rock barnacles,* followed by *blue mussels,* and finally *kelp* and other seaweeds. Seaweeds are covered in a separate chapter by Sylvia A. Earle. The balanoid-thallophyte biome follows a similar pattern as we move south, except that on the mid-Atlantic coast blue mussels are replaced by oysters and *hydroids* (colonies of tiny animals related to sea jellies), and sponges replace the kelp of the lower zones.

The lowest biome in terms of location is the *pelecypodannelid biome* (literally translated as clam and worm). It includes burrowing invertebrates, and these are least likely to be observed from above the surface.

SPONGES: Sponges are the most primitive of the *multicellular* (more than one cell) animals. They lack organ systems but contain specialized cells for digestion and movement of water. Water containing microscopic animals and plants, which are a sponge's food source, flows in through specialized openings called *ostia* and out through others known as *oscula.* The plankton trapped inside is digested. Sponges are sometimes hard to identify by casual observation. The encrusting *crumb-of-bread sponge,* which often has a greenish tinge, is common on pilings and rocks in New England. *Loosanoff's haliclona,* which is similar but more yellowish, is common further south to the mid-Atlantic region. The *red beard sponge* is one of the most colorful species and is found along the entire Atlantic coast and into the Gulf of Mexico to Texas. Its color is orange to bright red.

CNIDARIANS: The *cnidarians* (pronounced with the "c" silent and the accent on the second syllable) include the *sea jellies* (also called "jellyfish"), *corals, hydroids,* and *sea anemones.* Members of the group include some of the most delicate and striking of all marine invertebrates. Cnidarians come in two basic shapes—*polyps* and *medusae*—which provide the first step in identification. Corals. hydroids, and sea anemones are polyps. Sea jellies, which are medusae, are part of the plankton. Hydroids are small and delicate. To the naked eye they appear mostly as brown or yellowish hairlike masses. Actually, hydroids are colonies of polyps. Like most cnidarians, they are plankton-feeders. The only coral found in northern waters is *star coral,* which has a white polyp. The spectacular reef-building corals do not appear north of Florida waters, but these species ordinarily do not colonize pilings. On the Atlantic coast the most spectacular cnidarians visible to boaters are sea anemones. The most conspicuous is the *frilled anemone* found south to the mid-Atlantic region. Frilled anemones ordinarily are orange to orange-brown and sometimes grow to 100 millimeters (4 inches).

MOLLUSKS: The mollusks are a huge group of invertebrates. Briefly, mollusks are characterized by a *mantle,* which is a fold in the body wall that secretes the shell characteristic of the group, and by the *radula,* or filelike tongue. The radula has been modified by evolution to accommodate different feeding habits characteristic of individual species. Mollusks include the *chitons, snails, nudibranchs* (shell-less mollusks), and *bivalves* (e.g., clams, mussels, oysters). In northern areas the periwinkle is a commonly visible mollusk, as mentioned previously. Another common snail is the *oyster drill*, which grows to 25 millimeters (1 inch) and feeds by drilling holes in the shells of other mollusks with its specialized radula and sucking out the soft parts of its prey. Oyster and clam shells picked up on the beach often contain neat, circular holes left by oyster drills. Nudibranchs (literally naked gills) are among the most delicate and lovely of the mollusks. Nudibranchs of different species are found along the entire Atlantic coast. The *green nudibranch* is a common summer resident of south Florida, where it can be found on dock pilings in still water. The *rim-backed nudibranch* is found from Long Island Sound north. Nudibranchs are specialized feeders, often utilizing only a particular species of hydroid or other invertebrate. The predominant bivalves on pilings are blue mussels and oysters. As mentioned previously, blue mussels are northern animals and become less common in the mid-Atlantic region. There they are replaced by the *common oyster*, which in turn is replaced by the *coon oyster* in south Florida. Coon oysters, so named because they are utilized as a food source by raccoons, frequently grow attached to mangrove roots and are a common sight in marinas bordered by mangroves and salt creeks.

ARTHROPODS: The *arthropods* (the term means joint-footed) include the *crustaceans* (lobsters, crabs, and shrimps). Probably the most conspicuous crab south of Cape Cod is the *blue crab*, a large, aggressive species often caught in traps and handnets from docks. Blue crabs are opportunistic feeders, preying on mollusks, worms, and other invertebrates. Blue crabs sometimes grow to 225 millimeters (9 inches), as measured across the back between the tips of the longest spines. The *green crab* is smaller, growing to 150 millimeters (6 inches), and is not found south of New Jersey. *Fiddler crabs* live on mudflats or sandflats, mainly in salt marshes. They live in burrows in the mud, often in large colonies, emerging to feed at low tide. Male fiddler crabs are characterized by a greatly enlarged claw (either the right or left), which may be longer than the body is wide (body width averages about 38 millimeters, or 1.5 inches). Fiddler crabs range from Cape Cod south around the Florida peninsula to Texas. *Hermit crabs* inhabit the empty shells of marine snails. The *long-clawed hermit crab* is a common resident of inshore areas from Maine to Florida and the Gulf of Mexico. It commonly inhabits periwinkle and oyster drill shells.

No large, commercially valuable shrimps are common north of Virginia. Commercial shrimp fishing is a big industry in the Gulf of Mexico and the southern Florida Keys. Throughout south Florida, commercial shrimps are common around docks in winter when they can be dipped at night with handnets as they migrate through salt creeks. These shrimps grow to a length of 200 millimeters (8 inches).

The *American lobster* (sometimes called the "Maine lobster") can be found underneath docks in the northern part of its range (Maine to Virginia). South of Long Island Sound, American lobsters seldom are seen inshore. The *spiny lobster* is common in south Florida, where its long antennae are often seen protruding from under submerged rocks. Like their northern counterparts, spiny lobsters are opportunistic feeders, preying on mollusks, worms, and fish heads used to bait traps.

One of the most interesting "crabs" is not a crab at all but a relative of spiders and scorpions. This is the *horseshoe crab* found from Maine south to the Gulf of Mexico. Females grow to 600 millimeters (2 feet), including the tail spike; males are smaller. The spike is not venomous, contrary to what many believe. Adults migrate into shallow water in spring to lay eggs, when it is common to see a large female with a smaller male clamped firmly to the hind portion of her shell.

In northern and mid-portions of the Atlantic coast the predominant crustacean is the *barnacle,* of which several species are found. Although they resemble mollusks superficially, barnacles are crustaceans. A barnacle's shell consists of six plates that can be closed at low tide when the animal is exposed to air. When submerged at high tide, the barnacle extends its *cirri* to trap plankton. The cirri are feathery appendages that are moved in and out. An actively feeding barnacle extends and retracts its cirri several times a minute.

SEA STARS: Sea Stars or "starfish" are conspicuous because of their radial symmetry and *arms,* which usually are well defined and obvious. The most common sea star on the Atlantic coast is *Forbes' asterias,* found from Cape Cod south to the Gulf of Mexico. In the northern part of its range, Forbes' asterias inhabits pilings, where it feeds on mussels and barnacles. In the mid-Atlantic region, Forbes' asterias is a fierce predator of oysters and a menace to the commercial oyster industry. The *blood stars* are smaller and usually bright red. They range from Maine to Cape Hatteras, but are uncommon in shallow water in the southern portion of their range.

SUGGESTED READING

Guide to Identification of Marine and Estuarine Invertebrates: Cape Hatteras to the Bay of Fundy, by K. L. Gosner (New York: John Wiley and Sons, 1971).

A Field Guide to the Atlantic Seashore, by K. L. Gosner (Boston: Houghton Mifflin, 1979).

Dr. Stephen Spotte, a wildlife biologist, currently is director of Marinelife Aquarium in Mystic, Connecticut. He was born in Wheeling, West Virginia. Mr. Spotte has conducted field studies in the Bering Sea, Canadian arctic, coastal New England, and the Caribbean. He is the author of five books, and author or coauthor of more than sixty scientific and popular articles on marine biology and chemical oceanography.

SEASHELLS

Dr. R. Tucker Abbott

Man has hunted for seashells since he first took to the sea, whether it was a search for seafood, pearls, or shells for barter and ornaments. The ancients took to their boats when searching for murex shells from whence came royal Tyrian purple dye; and down through the ages seafarers have dreaded the ravages of the Teredo shipworm, a molluscan cousin of the clam.

Shelling today is a popular, worldwide hobby that delights the eye, gives the excitement of the search, and can sometimes bring profit to the lucky sheller—some shells may fetch several thousands of dollars. Even more rewarding are the many hours of study one may spend on the biological and educational aspects of conchology. Many new species are described each year, and the museum scientists welcome new information and fresh specimens for their scientific collections.

Shells are found from the arctic waters of Canada to the tropical reefs of the Gulf of Mexico. Each area has its unique molluscan fauna. In the north, the rock-dwelling periwinkles and the shallow-water Neptunes and *Buccinum* whelks abound, while in more southerly waters the colorful cowries, olive, and cones are commonly collected on sandy bottoms or in the rubble of coral reefs.

Mollusks are nocturnal and shun strong sunlight, hence collecting at night with the aid of a flashlight during a low tide can be very rewarding. Early morning or twilight hours in the evening, when the tide is just beginning to return, is the best time for collecting on sandbars and grassy flats.

Protected tidepools, crevices in rock jetties, and wooden piers are good collecting places. Most of the better shells are the live ones within reach of the snorkler or the scuba diver. Handpicking is the simplest, and you need only a cloth or plastic bag for holding your shells. Gloves and a short crowbar or diving knive are optional. After looking under a rock or coral slab for hiding mollusks, return the shelter to its original place in order to protect sea life from sunlight and wave action.

Scuba divers, because they can reach the habitats of mollusks at depths between 20 and 150 feet, are bringing in many beautiful species formerly thought to be rare. The best collecting is along the edges of the reef where there is an accumulation of dead coral and protective rock slabs. Fan the sand aside for burrowing snails, and feel along the surface of seaweed-covered rocks for rock shells. Take a sample of about a quart of shelly sand. Dried out later, and with the aid of a hand lens, you can find rare miniature shells.

Among the interesting mollusks found floating on the open sea, or associated with mats of sargassum seaweed, are the delicate Purple Sea Snails which produce their own floats of mucus bubbles; the white, inch-long *Spirula* shells from a deep-water squid; the minute brown Sargassum Weed Snail; and occasionally a Paper nautilus, or *Argonaut*. The latter are sometimes found in the stomachs of sailfish.

Boat owners have a great advantage in being able to reach out-of-the-way islands and reefs. They can dredge for shells and set traps or tangle nets. A small homemade metal dredge, with a mesh bag and with an entrance no more than eighteen inches wide and six inches high, can be dragged with a quarter-inch handline at depths of about fifteen to thirty feet of water. Greater depths or a larger dredge require motor-driven winches.

Do not overlook floating wood, beach-stranded sea fans, and seaweeds for tiny attached shells. When you catch a bottom-feeding fish, examine its stomach contents. Some rare shells have come from freshly caught fish. If you are in the area where lobster pots are repaired or where there are trash piles from scallop-shucking plants, you may find some quite remarkable abandoned shells.

Mollusks, like all shellfish, spoil rather quickly, and the meat must be removed before the stench becomes unwelcome. If you do not have freezer facilities available, boiling your catch for five to ten minutes in salt- or freshwater is the best method. Let the water cool slowly or shiny shells will lose their gloss. Pull out the meat with a bent pin or icepick, using an unwinding twist. Save the horny trapdoor, or operculum, if the snail had one. Snails and clams frozen for a couple of days, then thawed completely, usually can be easily pulled out of their shells. To store dried shells, wrap them in newspaper and pack in cardboard cartons. Be sure to write down the essential data on slips of paper labels—date and place of collecting, depth of water, name of collector, etc. In an emergency, small mollusks may be buried in cans or boxes of salt. The best preservative for any soft-bodied sea creature is 70 percent grain (drinkable) alcohol, or 50 percent isopropyl or wood alcohol (poisonous to drink). Formaldehyde (40 percent strength at the drugstore) can be brought down to about 8 percent by mixing one part to six parts of water. This solution will preserve fish but will ruin shells, unless each quart of 8 percent formalin is buffered with a tablespoon of baking soda. Tiny shells, soaked in alcohol for a week, can later be dried in the sun without further cleaning.

Shipboard is not an ideal permanent place for a shell collection, but small racks will hold the larger, colorful species. Bivalves and small snails can be mounted in picture mounts—shallow, flat boxes and a half-inch of cotton and a glass cover. Small shells can be stored in plastic tubes or snap-lid boxes, together with their locality data and identification slips. Shells may be mailed home, parcel post, in crunched newspaper and sturdy cartons if storage space is at a premium.

Identifying your shells can be an interesting challenge. There are many shell books on the market, some for beginners, others for serious conchologists. There are both worldwide and local identification guides, many of which may be in your home public library or university museum. Many areas have active shell clubs that welcome visitors and new members. Miami, Los Angeles, Honolulu, Suva, Sydney, Cape

Town, Boston, Chicago, and Washington, D.C., are only a few cities that have shell clubs. Three organization are willing to give you free advice:

Conchologists of America, 3235 N.E. 61st Ave.,
Portland, OR 97213

American Malacological Union, P.O. Box 394
Wrightsville Beach, NC 97213

American Malacologists, Inc., P.O. Box 2255
Melbourne, Fl 32901. (Offers free *Guide to Information on Shells and Mollusks.*

SUGGESTED READING

American Seashells, by R. Tucker Abbott (New York: Van Nostrand/Reinhold, (second edition) 1974). Hardbound, 3,000 species illustrated.

Seashells of North America, by R. Tucker Abbott (New York: Golden Press, 1968). Paperback, 850 U.S. species.

Seashells of the World, by R. Tucker Abbott (New York: Golden Press, 1962). Paperback, 600 worldwide species.

Compendium of Seashells, by R. Tucker Abbott and S. Peter Dance. (New York: E. P. Dutton, 1982). Hardbound, 4,200 color plates of worldwide shells, extensive bibliographies.

Standard Catalog of Shells, by R.J.L. Wagner and R. T. Abbot (Melbourne, Fla.: American Malacologists, 1978, supplements 1978, 1982). 15,000 species listed, values, world-size records, sheets for cataloging, maps.

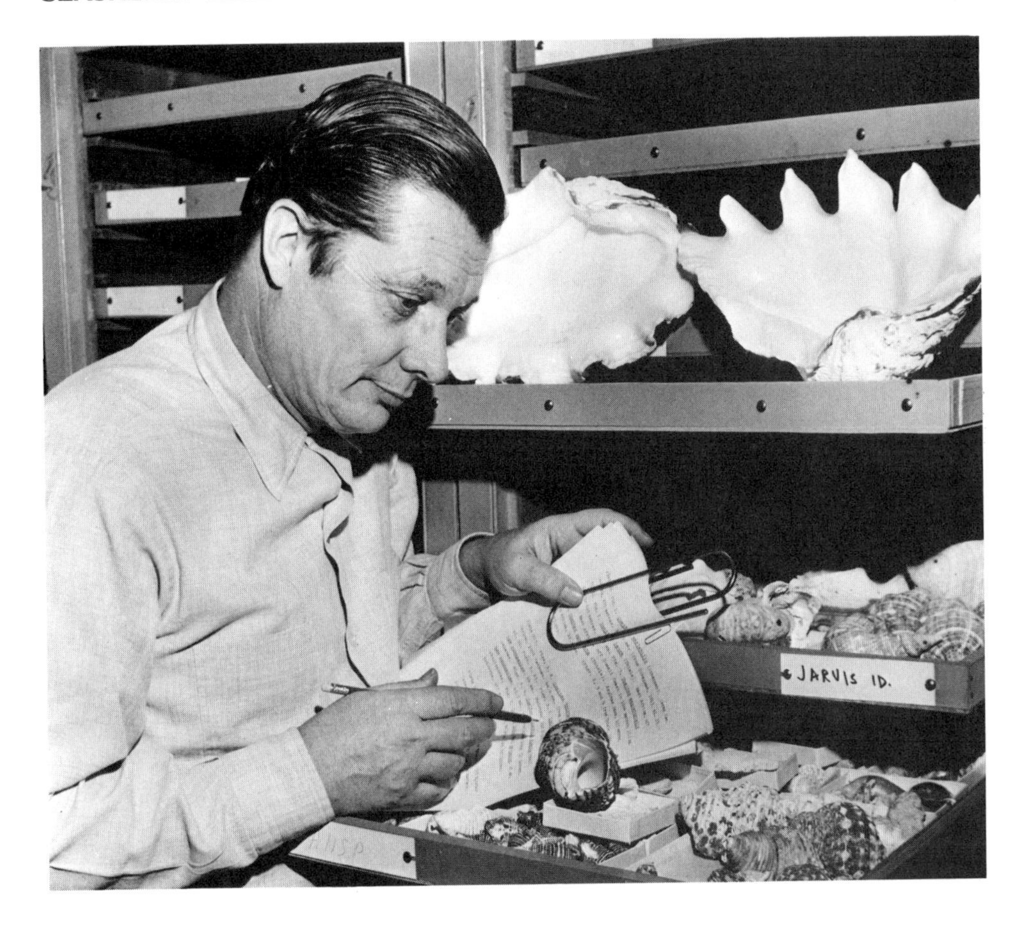

R. Tucker Abbott, author of two dozen books on shells, has spent the last forty years serving the science of malacology. Born in 1919 in Massachusetts, he attended Harvard University, served four years in the U.S. Navy in World War II, and then was a research scientist at such museums as the Smithsonian Institution and the Academy of Natural Sciences of Philadelphia. Dr. Abbott is editor of two scientific journals on mollusks, a former president of the American Malacological Union, and a trustee of the Bermuda Biological Station. Presently, he is president of the publishing firm, American Malacologists, Inc., in Melbourne, Florida.

SEAWEED

Dr. Sylvia A. Earle

	Love
Tamamo	Although no fisher
Ama to wa nashi ni	Reaping the gemlike seaweed,
Kimo kouru	I yearn for you
Va gu koromode no	So deeply that the salt spray
Kawaku toki naki	Never dries upon my sleeves

ANONYMOUS JAPANESE VERSE

Those who think about seaweed more often than not use words such as gumbo, scum, kelp, water moss, sea wrack, or simply "rubbish" to describe the growth on coastal rocks or the piles of marine plants that are cast ashore, sometimes in great quantities. Even more colorful names may be applied by those who find their boat's propeller fouled by a bit of weed or the hull covered by masses of enterprising plant pioneers

Mariners, however, have special opportunities that land-bound citizens do not to gain a different perspective about the plants that live in the sea. Given access to the ocean beyond the shore, even modest effort can result in a rather surprising and delightful new attitude about these beautiful, fascinating—but among the least appreciated—of the planet's inhabitants.

SEAGRASSES

Worldwide, about fifty kinds of plants that are lumped into the catch-all category of "seaweed" actually bear flowers, fruits, and seeds and have true roots and stems. They are known as seagrasses, plants that include the cool temperature eelgrass (*Zostera*) and warm temperate to tropical turtlegrass (*Thalassia*).

Generally, seagrasses occur in water less than sixty feet deep, except in very clear tropical seas where somewhat greater depths are attained by a few species. Great subsea meadows are formed by plants such as eelgrass. Seagrass beds along the shore of the coastal eastern United States are well known as "nursery" areas because they provide shelter and food for numerous young fish, crabs, shrimp, and other organisms. Scientifically, seagrasses are classed along with other flowering plants in the division Anthophyta. (Table I)

ALGAE

The great majority of seaweeds—the algae—do not have flowers, seeds, roots, or stems. In the broadest use of the term, all marine algae, even microscopic species, may be regarded as "seaweed," thus encompassing even planktonic diatoms and rock-encrusting coralline plants. Reproduction in the various groups of algae is accomplished by several kinds of spores, some involving the union of male and female cells and others that are asexual. Many can and do reproduce simply through fragmentation. In the macroscopic groups, the algal equivalent of stems are called "stalks" or "stipes" and attachment structures are known as "holdfasts" (sometimes embellished with hairlike rhizoids), and the leaflike structures are referred to as blades or fronds. The whole algal plant, to a marine botanist, is known as a "thallus."

Four major groups of algae make up most of the conspicuous, nonmicroscopic flora. Each group has a characteristic and distinctive assemblage of pigments that reflects its true biological affinities. Scientifically, it does not make sense to classify all yellow-flowered or red-flowered plants together, except the algae. The color of an alga is the first clue as to its identity. The groups (summarized in Table I) are as follows:

TABLE I

MARINE PLANTS

ALGAE	
Chlorophyta (Green Algae):	microscopic to massive plants, usually attached to a firm substrate.
Euglenophyta (Euglenoids):	Unicellular, sometimes free-swimming, sometimes living in mud or other soft substrate.
Chrysophyta (Diatoms, Golden-brown algae, coccolithophorids):	microscopic, planktonic plants.
Xantophyta (Vaucheria):	filamentous, microscopic mud-dwelling.
Pyrrophyta (Donoflagellates):	Microscopic, unicellular, planktonic plants.
Phaeophyta (Brown Algae):	microscopic to massive, usually attached to a firm substrate.
Rhodophyta (Red Algae):	Microscopic to massive, usually attached to a firm substrate.
Cyanophyta (Blue-green Algae):	Mostly microscopic, attached to firm substrate, but some planktonic.
FLOWERING PLANTS	
Anthophyta (Seagrasses):	Macroscopic, attached to rock or growing in mud or sand.

NUMBER OF SPECIES KNOWN

Algae	
Worldwide:	19,000
Marine:	about 15,000
Flowering Plants	
Worldwide:	250,000
Marine:	50

Blue-green Algae (Cyanophyta)

As the name suggests, these plants characteristically have a bluish-green hue (sometimes purplish to red). they are mostly small, filamentous, or unicullular, and most require microscopic examination to fully appreciate their intricate beauty. It may be difficult, at first glance, to convince someone that the masses of fine threads, spongy tufts, or slippery dark mats growing on rocks or boat bottoms are truly elegant in their form and structure, and are all the more to be appreciated because they are thought to be among the most ancient life forms known.

Relative to blue-greens, mankind is a latecomer, a new arrival on the planet. Their history goes back more than 3 billion years, whereas human ancestry is closer to 3 million years old.

One of the most intriguing aspects of blue-green algae is the ability of most to assume a wide range of forms, depending on where the plant happens to be growing. One particularly variable and cosmopolitan species of *Microcolous*, for example, lives in wet soil, fresh or salty water from Greenland to Antarctica, and from the floor of Death Valley to Pikes Peak. It had been given more than a dozen different names until a patient, astute botanist, Frances Dorset, determined that this plant of many faces—and many places—was in fact a single species.

Green Algae (Chlorophyta)

As the name suggests, Chlorophyta are conspicuously green. This is so because the green chlorophylls present in the cells predominate over other pigments. Some green algae are microscopic, and many occur only in freshwater or in moist habitats on land. About 1,000 marine species are known worldwide, with the greatest development occurring in tropical and subtropical seas.

One of the most well-known green algae is *Ulva lactuca,* sea lettuce, a plant of worldwide distribution that is regarded as tasty by a wide range of fish and invertebrate species, the marine iguanas of the Galapagos Islands, and even by some human beings.

In warm, clear seas, many green algae are calcareous. One kind, *Halimeda,* has more than fifty species, including some that contribute a major portion of the limy matrix to coral reef formations.

Some species have bizarre shapes, such as the warm-water green grape look-alike, *Caulerpa racemosa;* Mermaid's wineglass, *Acetabularia;* and Merman's shaving brush, *Penicillus.*

Brown Algae (Phaeophyta)

Like green algae, brown algae have chlorophyll, necessary for photosynthesis. The green is masked, however, by accessory yellow-brown pigments that help gather light as a part of the photosynthetic process—and give the "browns" their characteristic golden or dark brown color.

Included in this group are the largest and most conspicuous seaweeds of all. Certain giant kelp that grow in cool, Southern Hemisphere water have the distinction of being the largest plants in the world—as much as 300 feet long. Along the west coast of Northern America, one species, *Macrocystis pyrifera,* grows in forestlike stands that rival redwoods on land. Some grow as much as a foot a day and attain a length in excess of 150 feet.

Macrocystis is one of several large kelps valued for animal food, fertilizers, and for the production of a substance called algin acid that has numerous industrial applications. Paints, hand creams, toothpaste, chocolate milk, and ice creams are among the hundreds of products that may include this kelp derivative.

Along the eastern North American shore, brown algae do not reach such magnificent heights, but they are the largest and most abundant marine plants present in most rocky coastal areas. *Fucus,* sometimes called sea wrack, carpets intertidal rocks with sleek, glistening blades. Other browns live only on rocks below the tide line, some growing in stands more than six feet high.

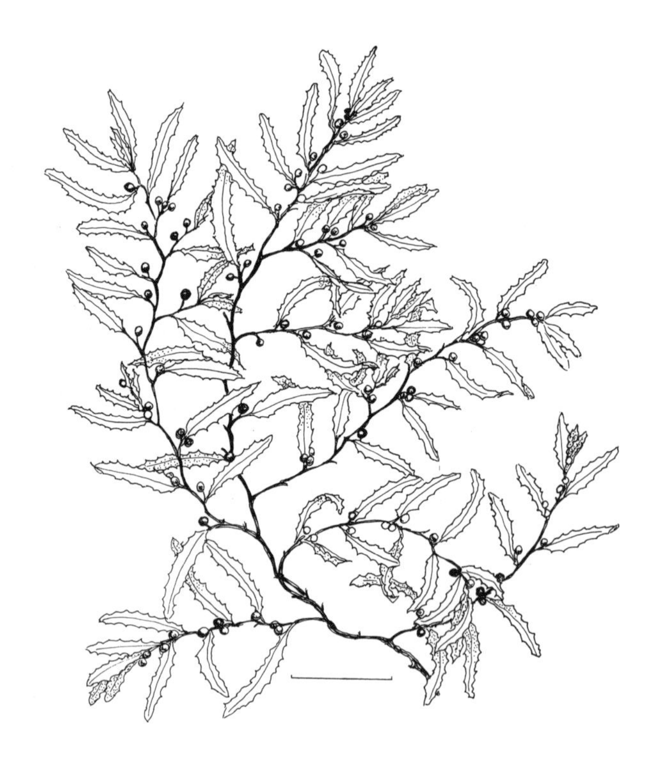

Sargassum Fluitans

In tropical and subtropical waters, the largest brown algae tend to be some variation on the theme of *Sargassum* or *Sargasso* weed. More than fifty species of *Sargassum* are known, mostly plants that attach to rocks and other firm surfaces. Two species are drifters, however. They do not develop holdfasts, but simply float about in large, planktonic rafts in the open sea.

Close inspection of drifting rafts of *Sargassum* illustrate the significance many marine plants have to animals in the sea. Shelter and food are provided for an amazing array of creatures as diverse as seahorses and flying fish to crabs, shrimp, and minute snails.

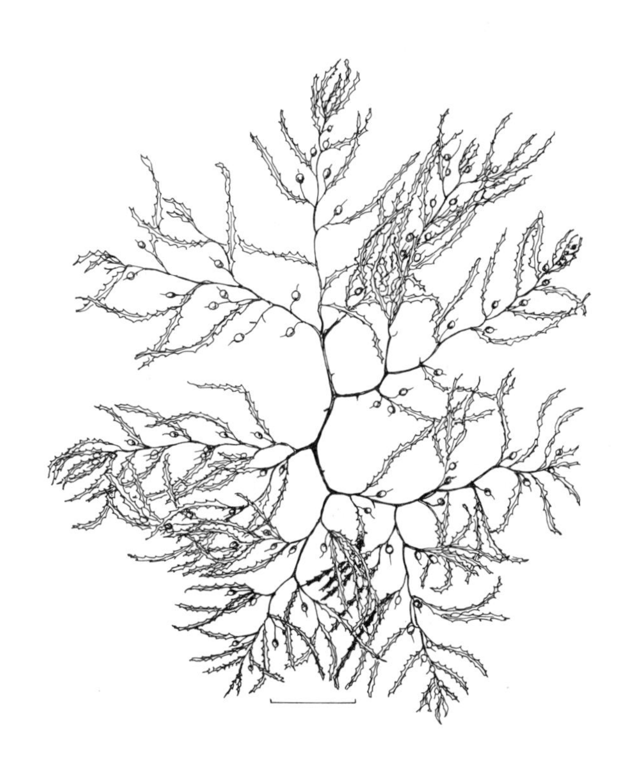

Sargassum Natans

Even in New England, drifting rafts of *Sargassum* may be cast ashore, brought far north of its usual range by the Gulf Stream. *Sargassum*, when viewed en masse, inspires descriptive terms such as those used in the opening paragraph. Individual plants, however, vie with Christmas holly or delicate ferns in terms of aesthetic appeal.

Red Algae (Rhodophyta)

Although not as large nor as conspicuously abundant as brown algae, red algae are the most diverse in terms of species. Warm-water "reds" tend to be smaller, often microscopic, but throughout the oceans of the world, the Rhodophyta outnumber the combined species of brown, blue-green, and green algae known.

As in all plants, green chlorophyll pigments are present in Rhodophyta, here by two accessory pigments, one bright red, the other bluish-red, and in such abundance that the chlorophyll is masked. The red pigments gather light energy and pass it to chlorophyll in the photosynthetic process. They also give the plants remarkably bright hues of pink, purple, red-tinged-with-green, and true red.

In tropical areas where coral reefs develop, certain encrusting, coralline red algae are of significance in binding together rock, sediment, coral, and other materials into a solid structure. The principal substance of some so-called coral reefs is, in fact, more plant than animal in origin.

Most red algae live below the low-tide line, and some occur to the edge of darkness in the deep sea beyond 600 feet. However, one of the best-known kinds, *Porphyra* (known variously as nori or laver or slook), is an intertidal dweller. Worldwide, people have come to value various species of *Porphyra* for food, either eaten straight from the rocks, or variously dried, salted, boiled in soups, wrapped around rice cakes, sprinkled on other foods as a condiment, served in salad, and even baked as laverbread.

Numerous other red algae are eaten (all are, in fact, edible, but some are more palatable than others). Dulse *(Palmeria palmata)* and

Irish moss *(Chondrus crispus)* are so highly prized for food in western Europe, eastern Canada, and the northeastern United States that they are commercially gathered and sold for use as vegetables, sweetened and flavored as a gelatin dessert (blancmange), otherwise used as an ingredient in various dishes. Irish moss (also called pearl moss, jelly moss, curly gristle moss) is the primary source of a gelatinous substance, carrageenan, that is used as a thickening agent in foods and for various industrial purposes.

To a marine botanist, the red algae have appeal that goes far beyond the palate. The intricacies of their life history, the diversity of species, the incredible variations in terms of structure and habit, the unusual pigmentation and related biochemical phenomena, the fact that they are simply elegantly beautiful have attracted numerous dedicated scientists to explore the unknowns of these remarkable plants. There is much yet to be discovered, even in the most basic terms, however. Since most red algae live beyond wading depths, many new marine species are being discovered for the first time through the use of scuba-diving techniques and small submersibles.

SEAWEED AS FOOD

People who have a long seaside history, particularly those from island nations, have come to appreciate and enjoy the culinary virtues of the plants that live in the sea. Hundreds of species have been relished for their differences in flavor, texture, and even beauty, much the same as vegetables on land have been throughout time. In Japan, it is not uncommon to have half a dozen or more species of seaweed served during a single meal.

There is no reason to be inhibited about tasting various kinds of seaweed fresh from the sea or cooked in imaginative ways. Unlike mushrooms and many land plants, no seaweed is deadly poisonous. One filamentous blue-green *(Lyngbya)* is reportedly toxic, but fortunately, its dark, stringy appearance would not be likely to tempt gustatory experimentation, anyway.

The significance of seaweed as food for humans is vastly overshadowed by the importance it has for the numerous turtles, fish, shrimp, snails, annelids, urchins, and other animals. As on land, plants in the sea transform energy from the sun through photosynthesis into sugar, and then into plant substance that forms the basic energy source for non-photosynthesizers such as animals and fungi. Without plants in the sea, the ocean would be a barren place.

Those who graze on macroscopic seaweeds include such well-known creatures as the California abalone. Without an assortment of red and green algae and tender young kelp plants to munch on, abalone could not survive, nor could the industry in California that is based on gathering abalone that in turn are served to human consumers.

Urchins of various sorts browse directly on seaweeds, as do limpets and numerous other mollusks. In cold water, invertebrate grazers dominate, but in warm seas, they have vigorous competition for their vegetable fare from large numbers of herbivorous fish. Parrotfish even have specialized teeth that enable them to nip off discrete bites of algae and to scrape hard rock surfaces to obtain encrusting species, as well as those that actually live embedded within coral. Large schools of surgeonfish travel like herds of grazing animals on land, browsing on seagrasses and algae as they move along.

Although rarely regarded as "seaweed," microscopic algae produce the vast quantities of food required to keep planktonic grazers such as copepods, krill, and numerous other minute animals well fed.

SUGGESTIONS

Those inspired to get to know seaweeds better might consider some of the following possibilities:

1. Get to know the plants on their own terms. Slip into scuba gear and glide through the fields, plains, and forests of the sea with the specific goal of seeing seaweeds, or at least try a face mask and flippers or a glass-bottomed bucket held over the side of a boat to

see what's going on below. The cast-ashore remains of algae and seagrasses bear as much resemblance to the living, growing communities of plants as a wood stack does to a forest.

2. Find out who and what they are. Recognizing that it is difficult to care about something until you know it in more than a superficial way, a brief list of references is appended that go further than this account and should lead to other good sources of information. One (Dawson, 1966) gives details about how to take individual plants and press them onto sheets of wet paper that are then dried, either for preparing permanent scientific reference collections or simply for lovely artistic renditions.

3. Pluck a few plants, selectively, so as not to disturb the nature of the area. To see the small creatures that live in association with the plants, float the seaweed in a white pan filled with seawater. Try using a hand lens or microscope to gain a proper perspective about the structure of the plant, as well as to better get to know the numerous microscopic animals that are inevitably present.

4. Seek out an expert. They are not as hard to find as one might think. Most coastal universities, museums, fisheries stations, and biological research institutions either have a resident authority or can provide direction. Marine botanists are typically surprised and usually delighted to discover someone who takes an interest in these widely unheralded but irresistibly appealing (once you get to know them) inhabitants of the ocean.

SUGGESTED READING

Marine Algae of California, by I. A. Abbott and G. J. Hollenberg (Stanford, Cal.: Stanford University Press, 1976).

The Plant Kingdoms, by H. C. Bold (Englewood Cliffs, N.J.: Prentice-Hall, 1970).

Introduction to the Algae, by H. C. Bold and Michael J. Wynne (Englewood Cliffs, N.J.: Prentice-Hall, 1978).

How to Know the Seaweeds, by E. Y. Dawson (Dubuque, Iowa: William C. Brown, 1956).

Marine Botany, by E. Y. Dawson (New York: Holt, Rinehart & Winston, 1966).

The Biology of Seaweeds, edited by C. S. Lobban and M. J. Wynne, Botanical Monographs, vol. 17 (Berkeley: University of California Press, 1981).

The Seavegetable Book, by J. Madlener (New York: Clarkson Potter, 1977).

Marine Algae of the Monterey Peninsula, by G. M. Smith (Stanford, Cal.: Stanford University Press, 1944).

Marine Algae of the Eastern Tropical and Subtropical Coasts of the Americas, by W. R. Taylor (Ann Arbor: University of Michigan Press, 1960).

The Marine Algae of the Northeastern Coast of North America, by W. R. Taylor (Ann Arbor: University of Michigan Press, 1972).

Dr. Sylvia A. Earle is a marine scientist with degrees from Florida State University and Duke University. She has served as fellow, research biologist, and curator at the California Academy of Sciences since 1976 and has been research associate at the University of California, Berkeley, and research fellow or associate at Harvard University since 1967. She has extensive field experience worldwide, including leading more than 40 exhibitions involving in excess of 4,500 hours underwater. She has used conventional and saturation diving techniques, submersibles, and the ADS systems, JIM, WASP, and MANTIS.

Dr. Earle is a member of several foundations, boards, and committees relating to oceanographic research, ocean policy, and global conservation, and has been given the Explorers' Club Lowell Thomas Award, U.S. Department of the Interior's Conservation Service Award, the Order of the Golden Ark by the Prince of the Netherlands, and named

Times Woman of the Year and the California Museum of Science and Industry's 1981 Woman Scientist of the Year. Since 1980, she has served on the President's National Advisory Committee on Oceans and Atmosphere.

WHAT'S GROWING on YOUR HULL: The FOULING COMMUNITY

Jack J. Rudloe

If you've been remiss and left your boat sitting for too many months in the water without hauling it out, you'll be introduced to a complex and dynamic assemblage of organisms known as the "fouling community." As the machinery hoists your boat up into the air, the bubbling and hissing of barnacles can be almost deafening. It is as if they know their world and all the miles they have traveled attached to your hull are coming to an end.

The noise seems to grow louder as the scrapers are raked back and forth, crushing the barnacles and shells and leaving round white basal scars on the wood, fiberglass, or metal. If the boat has been moored in a slip or remained at anchor for a year or longer, it may also blossom with luxurious mats of hydroids, mussels, and sea squirts. With a good beard of fouling growth, hull speed can be cut by as much as 40 percent and the fuel bill doubled. It is a constant battle with the sea, of pulling, scraping, and painting with antifoulant paints to keep a boat hull from turning into an artificial reef. Antifoulant paints help, but sooner or later the sea dissolves them away, and new life appears. Those with small boats that are hauled out on a trailer after each use needn't worry about the costs of the fouling community. They can sit back and enjoy its fascination.

Whenever any object, be it wood, steel, rubber, or plastic, is placed in the sea, it becomes overgrown with a diverse and easy-to-see assemblage of animals that varies tremendously in location and latitude. If your vessel is moored to a floating dock in warm brackish water, full of nutrients, then you can bet it will blossom within a few months. But it can sit at anchor in the crystal-clear waters of an ocean island, and chances are the water will be practically sterile, supporting few encrusting larvae.

Marina basins with sewage discharges are by far the best areas to create life on your boat bottom. The rich green soup of diatoms and phytoplankton sustains the seed stock.

It all begins with a primordial slime. Even freshly sunk wooden wharf pilings that leach out noxious repelling creosote, designed to destroy all life, are eventually conquered by sea slime made of bacteria and later filamentous algae. This forerunner of new life lays down a coating and protects the first pioneering animals from man's poisons. Usually the first to settle are the barnacle larvae, that have a great propensity for fiberglass boat bottoms, or if there is a heavy set of oyster larvae drifting by when an object is first submerged and sufficiently coated, those will be the primary settlers. Then come the sponges, sea squirts, hydroids, etc. The community may change drastically with the seasons, particularly frigid-temperature waters where blistering cold winters will kill off all but the most rugged survivors.

Some years wharf pilings, styrofoam floating docks, and boat bottoms become heavily encrusted with certain sea squirts; other years they support rich growths of fluffy pink hydroids and brown tufted bryozoans, or thick scalps of mussels. Salinity greatly affects what settles. A year of drought makes the water saltier, allowing a more diverse community to grow. A very wet year will have fewer animals. Animals also vary tremendously from open exposed coasts to protected harbors, finger-fill canals, and dredged-out boat basins where there is seldom any wave action.

The longer an object sits in the water, unmoving and stationary, the more life grows on it. A floating dock, a wharf piling, a bridge abutment, a sunken automobile, or a more or less permanently moored vessel becomes an artificial reef. Animals beget animals.

Eventually there are two categories of life within the fouling community—the sessile and the motile, the sedentary and the crawlers. The sessile forms, such as barnacles, oysters, and serpulid worms, cement themselves to a hull or piling by calcareous secretions. Mussels use specialized byssus threads, while sponges, hydroids, bryozoans; sea squirts fuse their living tissues directly onto the substrate.

These rely upon the movement of seawater, the tide, currents, and winds pushing water to provide them with their food. Plankton, the microscopic plants and animals that float in the water, and tiny particles of detritus from the marshlands and mangroves make up the feast.

The mechanisms of collecting food from this soup of life are similar, whether it's the feathery legs of barnacles that jump out of their protected shells like jack-in-the-boxes to snatch a passing copepod, or the hairy sweep of a bryozoan zooid. Hydroids and anemones reach up with inviting colorful tentacles and sting the passersby to death, while filler feeding sponges, oysters, mussels, and sea squirts pump water through their entire bodies straining out food, receiving oxygen, and passing out wastes.

Within such a community there is not much competition for food, but a tremendous competition for space. The weaker are crowded out, the stronger flourish. And from the moment the larvae or young are scattered into the sea they must find holdfasts or perish.

Then there are the hunters within the community, the ones that prowl the forests of hydroids and hide within the crannies of barnacles, or the mounded clusters of leathery sea squirts. These are the errant polychaete worms and nudibranches, small shelless snails. Hundreds—if not thousands—of small mud crabs and sea fleas or amphipods spend their lifetimes creeping among the colonies, picking out worms, eating decaying sea squirts or pieces of sponges. Or perhaps they feast on tiny snails or newly attached small bivalves.

Hairy brittlestars wave their snaky legs out from the canals of sponges. Tiny dragonlike nudibranches creep and slide over the hydroid colonies, devouring their pink polyps and transferring the hydroid's stinging cells into their own bodies, protecting them from hungry fish. Uncountable millions of tiny monsters called skeleton shrimp bob their peculiar elongated bodies as they crawl from branch to branch of the hydroids, scraping off diatoms and unicellular algae. A bristleworm seizes an unsuspecting sedentary worm in its powerful jaws. Slipper limpets creep slowly along, grazing on the surface like cows.

Meanwhile, myriads of tiny copepods orbit the fouling community, feeding on waste that provides food to passing fish. Smaller fish, such as gobies and blennies, and transparent grass shrimp hide within the fouling community, seeking refuge from their predators.

All of this life is fascinating to watch. Simply tear off clumps of fouling growth, spread it out in a jar or a glass dish, and examine it with a magnifying glass or a dissecting microscope.

But if your vessel is made of wood and hasn't been pulled for a year or two, some of these creaturres can be deadly. It's the ones that get *into* the wood and tunnel their way through your hull that can be devastating. When all the barnacles and other fouling growth are scraped away, and you see tiny pinholes staring back at you, you have met the wood borers. It can be a costly encounter.

Shipworms and gribbles have sent many a ship to the bottom after honeycombing the hull with their burrows. Shipworms, the best known of which are *Toredos,* are not worms at all, but very primitive clams modified with teeth that rasp away at the wood and twist and turn until their burrows are made. They do not eat the wood, merely prepare it for protected, comfortable homes. Their shell is only a tiny portion of the clam's great elongated body, which may stretch for a foot or more through the wood out to the surface, where it can siphon in water and remove plankton. Perhaps to make their burrows more comfortable or to lend support, they secrete a lime coating.

If you break apart an infected board, you will see dozens of white tubes and a large fleshy "worm" inside. Often the wood is so weakened that a two-by-four crumples in your fingers. When shipworms die, they leave sand-filled tubes behind and can destroy the stoutest timbers.

But even more ominous to the wooden boat owner are the gribbles that can actually digest cellulose and devour wood. Tiny crustaceans called isopods, gribbles are related to pill bugs, or rolly-pollys, that one finds under rocks in gardens. When you see a wharf piling that is eaten down to a point, as if someone had put it into a gigantic pencil sharpener, then you're looking at their handiwork. They are scarcely two millimeters long, but they have caused mighty bridges to crumple into the sea. When you walk out on a dock and it wobbles dangerously, or dig your pocket knife into the keel and it goes through like peanut butter, then beware the gribbles.

They are practically impossible to see with the naked eye, but cut off a chunk of wood, examine it under a dissecting scope, and you can watch them in their own little burrows eating away. They eat their dry weight in wood every ten days, which puts termites to shame. Each has a highly developed pair of jaws; the right one with a rasp and the left with a file. As they saw away at the wood, fragments are crushed and swallowed. They are amazingly resistant to antifoulant paint. All they need is one little scraped area where the surface is exposed, and they move in and take over.

Nevertheless, the requirements for gribble homesteading are not simple. Gregarious creatures, they prefer to live in wood that is already infested with others of their kind. If an isopod lands on a surface and eats a burrow, and after a while no others come to stay, it will leave and go elsewhere to find company. Sometimes they will abandon wood that has been almost devoured and move on.

Usually this migratory movement takes place at night. The males leave first, and when they are established in a new piece of wood, the females join them. They arrive swimming, and crawl over the surface, going from burrow to burrow to see if a male is there, poking their antennae down inside. If the burrow is occupied by the opposite sex, they disappear into their new home and raise a family. Between the borers, fungus, bacteria, and all the encrusting creatures that attach, it is only a matter of time before wood disappears entirely.

As a biologist, a boat and dock owner, I have very mixed feelings about wood borers. Costly though they are to humans, they have their place in the design of nature.

They clean the rivers and estuaries of wood. Rivers help nourish the sea by dumping leaves, bark stems, and branches of upland vegetation into these arteries of water that feed out to the sea. But they also carry down huge logs and often entire trees that are washed away by currents slicing into the riverbanks. Sooner or later—it may take years—it all ends up in the sea. So do pines and palms, and mighty oaks along the forested shorelines when a hurricane sends gigantic waves pounding the shore. Foundations of houses are periodically washed out during storms and topple into the water. Who cleans up? Busy little borers, the bivalves and isopods digging their burrows into wood. When you hear the sick "clang" of a boat propeller colliding with a submerged floating log, you have wood borers to thank that it doesn't happen more often.

A dead tree carried out into saline water eventually becomes waterlogged and sinks. Then it blossoms into a flowering garden of red beard sponges, purple and pink glistening chunks of sea pork, waving clusters of pink hydroids and white sea anemones. It becomes a gathering place for schools of fish—grunts, spadefish, sea bass, mullet, snapper, and grouper in warmer latitudes. They all come to pick and choose among the abundance.

A sunken boat on a barren mud bottom becomes the richest habitat around, a gathering ground for fish, often a home for lobsters. Man, wanting to increase the fishing grounds, is now deliberately creating artificial reefs by sinking cars, scuttling obsolete ships, and hauling bargeloads of junked cars, concrete blocks, culvert pipes, and rubber tires out and dumping them.

When he puts out sea buoys that mournfully clang their warning to ships, he creates yet another habitat. A simple scraping with a dip net on a channel marker produces a profusion of life, millions of tube-building amphipods, flatworms, polychaetes, and minute crabs, so many that it boggles the mind. The life is ephemeral; it reproduces explosively, in incredible numbers, before it dies off, or is devoured by fish. And the fish are eaten by man.

The fouling community is an expense and a nuisance to the boat owner who must scrape the barnacles and other growth off his hull. But when he casts his line near a buoy and reels in a bluefish that has been feeding on shrimp that have been feeding on amphipods and worms, the fouling growth is an asset. And if he scoops up some of this life and examines it closely, it becomes a fascination and wonder indeed.

SUGGESTED READING

Bibliography of Protozoa, Sponges, Coelenterata, and Worms, by D'Arcy W. Thompson (Boston: Longwood Press, 1977, reprint of 1885 edition).

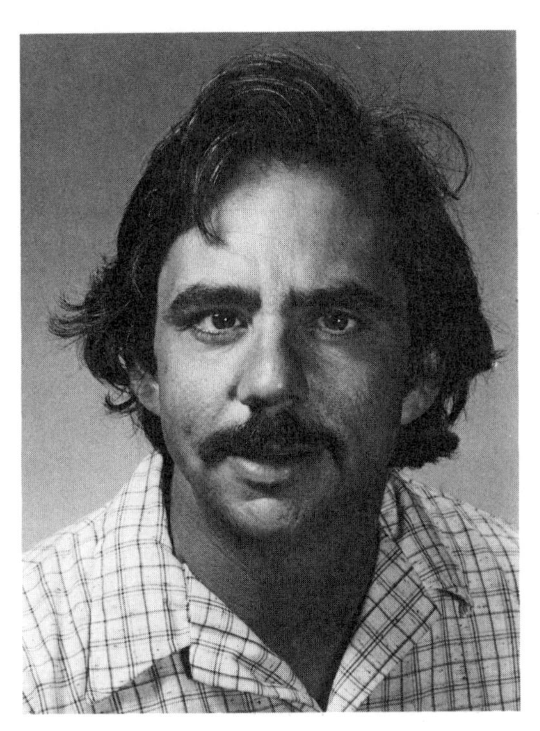

Jack J. Rudloe is the president of Gulf Specimen Company, Inc., in Panacea, Florida, a supplier of living marine fishes and invertebrates for

teaching and research. He is the author of eight books and currently is writing a ninth book, his first novel. He has written nearly twenty articles for Audubon, Sports Illustrated, National Geographic, Scientific American, *and several other publications.* The New York Times, The Wall Street Journal, Sports Illustrated, *and* Audubon *among many others, have carried features on Mr. Rudloe's life as a marine collector and his efforts to save Florida's marshes.*

Mr. Rudloe has traveled to Madagascar, Honduras, Nicaragua, Haiti, Costa Rica, Guatemala, and Suriname in search of such creatures as the giant isopod, deep-water sharks, and the giant toadfish.

PLANKTON

Dr. George D. Ruggieri

The first time I saw it, I couldn't believe my eyes. It was huge, much bigger than the largest whale. It undulated at the surface, rolling with the motion of the sea, glimmering, almost rubylike. It was a red tide made up of billions of one-celled organisms, one part of the world of plankton.

The oceans can be conveniently divided into the pelagic and benthic realms. The benthos is the sea bottom, and the organisms found there either burrow into it or reside upon it. Clams, oysters, polychaete worms, sea stars, sea cucumbers, etc., are representative of the benthos. The pelagic zone is the seawater itself; and the organisms that reside there either float about or actively swim. Nekton (Greek, *nektos:* swimming) is the term encompassing all those animals capable of swimming at speeds sufficient to advance against the current; some of these active swimmers even undertake extensive migrations. This group comprises squid, cuttlefish, seals, whales, dolphins, turtles, tunas, swordfish, sharks, and most adult fish such as mossbunkers and bluefish. The plankton (Greek, *planktos:* wandering) is comprised of the floaters or drifters. Many of these do indeed have organelles for locomotion, but they are weak swimmers whose mobility is geared toward vertical movements; their horizontal movements are at the whims of the ocean currents. A great many organisms are drifters; the plants among them are referred to as phytoplankton, and the zooplankton are the animals. Those organisms that spend their entire lives as members of the plankton are referred to as holoplankton. All phytoplankters and certain of the zooplankton, such as copepods, jellyfish, and comb jellies, are permanent residents of the planktonic realm. Many other animals are temporary visitors to the plankton world. These are referred to as meroplankton; they leave the plankton at early stages in their life cycles. These include larvae (young stages) of most bottom dwellers such as clams, oysters, polychaete worms, sea stars, sea cucumbers, etc. Even young stages of many fish comprise parts of the transient plankton till they are large and strong enough to escape the horizontal pull of the water.

SEA PASTURES: PHYTOPLANKTON

These floating microscopic plants contain chlorophyll, and, therefore, to grow and reproduce they need water, sunlight, carbon dioxide, and certain inorganic nutrients such as phosphates and nitrates. The water and CO_2 are always abundant, so the availability and interplay of solar energy and inorganic nutrients determine the abundance of phytoplankton. In temperate waters, such as those along our northeast coast, there is a burst of phytoplanktonic growth in spring. The cold waters of winter have brought up to the surface the rich nutrients from the bottom, and these nutrients, coupled with the warming sun, lead to a rich, lush growth of microscopic plants. This growth of plant life or the primary production of new organic material (protein, fats, and carbohydrates) is the first and essential link in all the food chains in the sea. Herbivorous zooplankton feast upon the phytoplankton, and they in turn are devoured by carnivorous zooplankton, and they by larger predators (little fish are eaten by bigger fish!). The chief difference between primary production in the sea and that on land is that the phytoplankton is eaten almost entirely by zooplankton, whereas on land comparatively little plant life is eaten by herbivores. On land, the grasses, bushes, and trees are long-lived; in the sea these primary producers are short-lived. The phytoplankton are literally here today and gone tomorrow—even those that aren't eaten!

Most invertebrates and fish spawn in spring to take advantage of the abundance of planktonic life that results from the bloom of plant life.The phyto- and zooplanktonic organisms supply the food to sustain the young stages of invertebrates and fish, and enable them to grow to adulthood. The cycle wanes during the summer as the nutrients are used up and plant growth diminishes; in temperate waters, another shorter and weaker spurt of plant growth occurs in autumn.

In polar waters one magnificent burst of phytoplanktonic growth occurs in midsummer, and this allows the zooplankton, notably krill, to abound and thus provide the food for the great baleen whales. In the tropics—except in areas of upwelling, i.e., where bottom cold water is pushed to the surface, bringing with it the rich nutrients necessary for plant growth—phytoplanktonic growth is low throughout the year. Oceanic waters with a low level of productivity are clear and blue. The paucity of planktonic organisms allows light to penetrate to a considerable depth, and the blue wavelengths are reflected from the water molecules. The color of seawater rich in plankton is green. Plankton and organic detritus reflect yellow wavelengths which, combined with the blue wavelengths reflected by the water molecules, produce a green color.

The role of bacteria in the sea is still being unraveled. But like certain bacteria on land, some species are active in the decomposition of organic matter, providing nutrients for plant growth, and they also utilize organic material for their own sustenance. The bacteria themselves are then consumed by marine animals, from protozoa to organisms higher up in the food chain.

The principal phytoplanktonic organisms are microflagellates, diatoms, dinoflagellates, and coccolithophores. Diatoms are important first links in the food chain. They are the main food source for large swarms of copepods, which in turn are the primary food of small fish such as herring. As food producers, dinoflagellates are not on a par with diatoms, because many species of dinoflagellates have a cellulose shell which is indigestible and others have long spines which make them difficult to ingest. Dinoflagellates are the organisms most frequently responsible for the phenomenon known as red tide. When the red tide organism contains a toxin, it can result in massive fish kills. An outbreak in 1957 in Tampa Bay, Florida, in which millions of fish were killed, had concentrations of 100 million organisms per liter of water. That's 100,000 dinoflagellates in one cc (more than 4,000 per drop)! Other toxin-containing dinoflagellates, when concentrated in shellfish such as clams or mussels, will lead to poisonings in humans who eat such shellfish. It is now known that ciguatera poisoning, which affects fish in the Pacific and Caribbean, has its origin in a toxic dinoflagellate. People can be poisoned by eating the flesh of ciguatoxic fish.

ZOOPLANKTON: DRIFTING ANIMALS

Holoplankton—the permanent members:

Copepods are the most dominant of all the zooplankton, and the herbivores species are small animals, probably no larger than five mm in length. Copepods of the genus Calanus probably are the most numerous animals in the world and are the most important single food animal in the sea. Egg production and early development of different copepods are geared to take advantage of the seasonal changes in plant productivity. For example, the same genus of copepod will produce three generations per year in one locale; in other locations two generations and one generation, all apparently responsive to phytoplanktonic abundance. In polar waters, krill, which are shrimplike zooplankton, occur in dense swarms and constitute the principal food of whalebone whales.

Although the crustaceans (copepods, krill, shrimps, etc.) form the bulk of the permanent zooplankton, representatives from other animal phyla also grow to maturity and breed as permanent members of the marine zooplankton. Comb jellies and jellyfish, some of which are quite large, are permanent inhabitants of the planktonic realm, as are certain species of snails and arrow worms.

Meroplankton: temporary residents of the zooplankton

Every major phylum of the animal kingdom is present as transient members of the zooplankton. The larvae, or young stages, of animals as diverse as sponges, flatworms, barnacles, crabs, clams, oysters, mussels, sea urchins, sea stars, sea cucumbers, and various fishes are all occasional visitors to the world of plankton.

Planktonic organisms are often useful in identifying natural regions in the oceans; these are referred to as indicator species. Abundance of diatoms are characteristic of polar waters, and dinoflagellates, coccolithophoriids, and blue-green algae are indicative of more tropical waters. The presence of a certain species of arrow worm is indicative of a specific water mass. But the abundance of another arrow worm of the same genus signals a different water mass.

The world of plankton, though consisting primarily of tiny organisms, is at the base of every food chain in the sea. The following are examples of food chains in the sea:

1. Phytoplankton → Anchoveta
2. Phytoplankton → Zooplankton (*Euphausia superba*, krill) → Baleen whales
3. Phytoplankton → Zooplankton (copepods, etc.) → Planktivorous → animals (e.g., herring, mackerel) → Fish-eating fish (e.g., bluefish, sharks, etc.)
4. Phytoplankton → Benthic herbivores (clams, mussels, polychaete worms, etc.) → Benthic carnivores → (plaice, cod, etc.)

→ Fish-eating fish

5. Phytoplankton (esp. small flagellates) → Microzooplankton (herbivorous protozoa) → Macrozooplankton → (carnivorous crustacea)

→ Megazooplankton (chaetognaths, Euphausids, etc.) → Planktivores (lanternfish, saury) → Oceanic carnivores (squid, tuna, etc.)

Crustacean copepods are the most abundant of the zooplanktonic organisms. And many copepod species perform the essential function of converting vegetable matter of the phytoplankton into animal protein. These small copepods are the herbivores of the sea; their relationship to the phytoplankton is the same as that of cows to grass on land. The zooplankton then is the bridge between the microscopic pastures of the sea and the largest marine creatures. Copepods and euphausids (krill) are so efficient at utilizing phytoplankton and exist in such immense numbers that their importance in the economy of the sea cannot be overemphasized. Copepods constitute the largest single group of protein producers in the *entire* world. For example, a single pair of a small copepod *(Tisbe furcata),* having a volume of only 0.14 cubic millimeters, produces four generations in 100 days. If all the offspring survived and also propagated within the same period, the first pair would have given rise to 1,055,000,000 individuals with a total volume of 70 liters.

Many planktonic organisms are being studied in laboratories. They are being grown in pure culture to evaluate their role in the sea, to exploit their use in aquaculture, and to learn more about the effects of pollution on their growth and reproduction. Certain toxic dinoflagellates are being studied in an effort to isolate and chemically characterize their toxins. Some of these toxins act so specifically on excitable tissues such as nerves and muscles that they are being used as molecular probes to learn more about how nerves and muscles function.

The best way to obtain a sample of plankton is to tow a cone-shaped net (of bolting cloth, silk, or monofilament nylon) through the water. The wider end of the net is kept open by a metal hoop, and this is attached by rope bridles to the tow line. The narrow end is closed off by a receiving jar or bottle. The mesh size of the plankton net for obtaining representatives of phyto- and zooplanktonic organisms should be 150–180 meshes per inch. The best towing speed is about 2 nautical miles per hour.

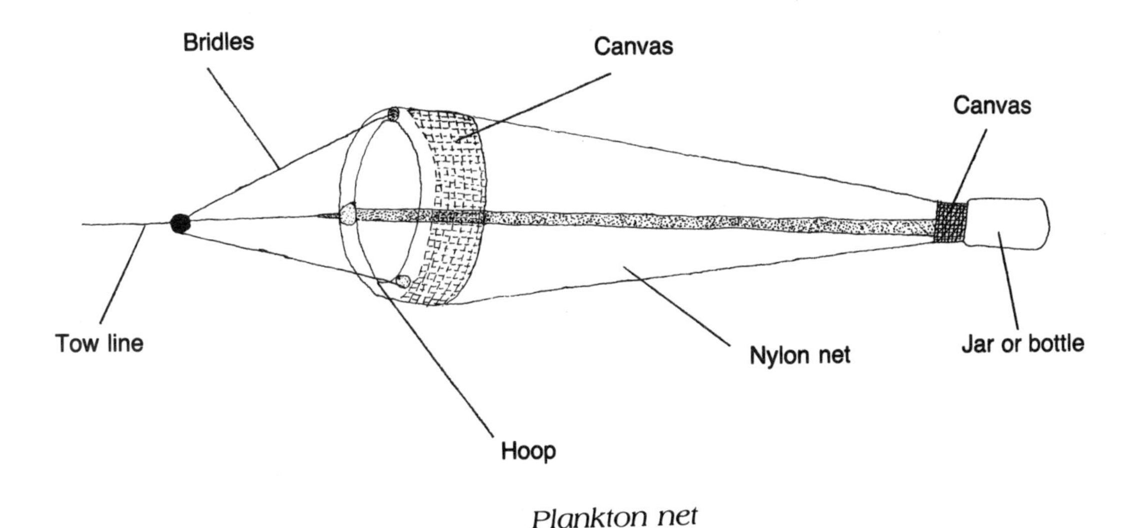

Plankton net

Individual members of the plankton are usually too small to see with the unaided eye, but large numbers of them in a plankton tow can be appreciated. But what should be most appreciated, whether one sees them or not, is the importance of planktonic organisms to the well-being of all life in the oceans.

SUGGESTED READING

Marine and Fresh Water Plankton, by Charles C. Davis (East Lansing: Michigan State University Press, 1955).

Marine Plankton: A Practical Guide, by G. E. Newell, and R. C. Newell (Highlands, N.J.: Humanities Press, 1966).

Dr. George D. Ruggieri, S.J., is director of the New York Aquarium and Osborn Laboratories on Marine Sciences. He is a member of the executive committee of the board of trustees of Fordham University and is an adjunct professor at both Fordham and New York University. Twice co-chairman of the Food and Drugs From the Sea Symposium, he co-authored The Healing Sea. *In 1977 he received the Alumni Merit Award from St. Louis University and an honorary doctor of science degree from St. Joseph's University in 1981. A member of the Society of Jesus, Dr. Ruggieri has complemented his career as a a marine biologist with his vocation as a priest.*

WHERE FRESH MEETS SALT: ESTUARINE ECOLOGY

Dr. Kent Mountford

All boaters harbor misty dreams, sometimes realized, of real bluewater sailing—the incredible blue of Gulf Stream and Bahamas, porpoise running under the bows—but reality is that most of us do our boating, most of the time, in estuaries: those coastal inlets, bays, and river systems that offer us such variety and as often the challenge and frustration of boating close to the bottom, close to shore, and close to other boats!

What is an estuary? I hesitate to define because we scientists can argue the matter and its nuances for days, but, just for you, I'll be cavalier and state there are only two considerations: that a body of water be semienclosed (that is, retaining free interchange with the sea) and that the water inside be partly diluted by freshwater.

On the East Coast of North America, furthermore, I'll suggest there are just two basic types of estuary, although the gradations between them are almost infinite . . . exactly what makes them interesting to us mariners.

- RIVER-MOUTH ESTUARIES range from the craggy bluffs and ferocious tides of the Bay of Fundy in Nova Scotia to the mighty Chesapeake, a classic drowned river valley, slowly submerged (and still being submerged at perhaps 4 mm a year) since the end of the ice age 10,000 years ago.

- BAR OR BARRIER-BUILT ESTUARIES are more distinctly enclosed, by wave-built islands or peninsulas. These are usually of sand, like those bordering Long Island's south, Atlantic-facing shore, but they can also be of cobbles and gravel, like the spit at Scituate, Massachusetts; or even of coral and coral rubble like those off Biscayne Bay, Florida.

The principle is the same: Restrict access to the ever eager sea, add freshwater at the landward end, and you have an estuary. In the pristine state, these coastal systems are incredibly rich in natural resources. In our time they've been the deposit sites for much of our pollution. Even so, estuaries have always been stressful to the special creatures that can survive their harsh requirements for life.

Harsh, you say? Why, one morning last June you were anchored in some placid estuary, with gulls wheeling, clear water beckoning you for a swim, and the water's surface just ruffled enough to intensify the blue it reflected from a cloudless sky.

Yes, harsh. The estuary is a place where many forces come together, and it is thus a place of radical change, almost continual change, in temperature and salinity. In the north, they say: "There're just two months, July . . . and winter." In the mid-Atlantic region, tremendous swings from wholesale freezes in winter to the slow boil in sultry soup of summer. We humans survive the seasons but only with a great deal of support equipment. Organisms in the estuary feel every degree change, every day of the week.

The struggle often includes tremendous and rapid changes in salinity, the amount of available oxygen, and (with tides) even the absence of water, with attendant exposure to drying sun and wind.

Remember, when we were kids, how we used to stay in the water so long we came out with our hands wrinkled? (Not to mention with blue lips because we'd gotten so cold!) Our hands and feet, especially, wrinkle that way because of the loss of free water in our bodies. Our fluids are flowing out, trying to dilute the whole sea and make it the same concentration as fluid inside our bodies. The same thing happens to creatures in the estuary, except that they have a biological pump, a system called "osmoregulation" that pumps salt out or in to maintain their body fluid balance. Like any pump, it takes energy to operate, and, for the organisms, it's quite like us climbing a long continual hill.

Among all the world's creatures, there are relatively few (OK, admittedly thousands . . . but among a million) that can manage all these stresses at once. If you were to don scuba gear and swim slowly over a mile of coastal or tropic ocean bottom, chances are you'd see hundreds of different kinds of animal (and possibly plant) life, but few numbers of each as a rule. In the estuary, probing the mud and waters carefully, you'd see many fewer kinds, but, at least for some, vast abundance. In the estuary, those fewer kinds that can survive often encounter conditions to which they're about perfectly suited; conditions where their competitors or predators, those who would eat them can't flourish. The result of this happenstance is what we've often, and mistakenly, assumed to be the estuary's inexhaustable abundance. In colonial times, the Chesapeake had immense "reefs" of oysters, sheltered from predatory snails and boring sponges by lower salinity, fed by healthy plankton diatoms growing on nutrients from the tributary rivers and their rich, virgin forests. Oysters abounded for centuries, too deep for Indian or colonist to gather until man invented tongs, rakes, and dredges to harvest a resource that had been untouched for 6,000–8,000 years.

Similar scenarios can be developed for other shellfish of interest to man: scallops in the bays, soft clams on New England tideflats, hard clams in the barrier-built "lagoons." The story could be told yet again for many of the fish that find safe haven, utilizing special conditions in the estuary, during parts of their life cycle when they're especially vulnerable.

Our valuable striped bass populations (called rockfish in the Chesapeake and nowhere else in the world) spawn each spring in tidal freshwater portions of our river mouth estuaries. In the critical few days when the hatched fry have consumed all the available nutrients in their yolk sac, and while their tiny mouthparts are still developing to where they can feed normally, they rely for survival on a critical mix of the right temperature, the right amount of river flow, the right concentration of nutrients, and just the proper amount of plankton growth. In a good year, only a few of thousands survive to adulthood anyway, and this is enough to produce harvests in the tens of millions of pounds. A change in the survival rate, even out in the third decimal place, makes the difference between banner and miserable years. Impose man's many complex polluting systems on that delicate equation, and the linkage between the valued resource and our lack of gentleness with the living system becomes apparent.

Life has a way of circumventing many of the obstructions nature puts in its way. Given enough time, it will find its way around almost anything. Enough time is the key; and we've added stresses and pollutants so quickly in the last few decades that the living system can't adapt. The valuable resources in most of our estuaries are suffering.

Most of us become intimately familiar with one or two estuaries, the ones where we do our boating. There's nothing wrong with that at all, but take the opportunity to really learn something. Study your charts and imagine how the tide sweeps in from the sea. How has this shaped the channels you navigate; how has man had to alter nature by dredging channels to give you access? In your special area, have the saltmarshes fringing the edges been destroyed for housing developments, honeycombed with residential lagoons that lie stagnant for months without ever seeing enough flow to change the water?

Where does the freshwater that dilutes the sea come from? Is there a mighty river like the Hudson or Susquehanna? Or are there only shallow sluggish streams that don't seem to flow enough even to account for the reduced salinity you know is there. Barnegat Bay in New Jersey is such a case, and the state geologist, Kemble Widmer, years ago told me more water actually entered the bay through the sandy western shore bottom, as percolating groundwater from the Jersey Pine Barrens, than entered by streamflow. He showed me a high-altitude photograph taken one winter when the bay was ice-covered. Along the western shore, the relatively warm groundwater, entering under the ice, had melted the entire ice mass clear of the beach. Thus detached, it had drifted on a strong northwest wind across to the barrier island

and could be seen there, packing up on the beach. I later verified personally that the ice had not only mounded up to twelve feet high as grounded icebergs in the shallows, but along miles of island shoreland, it had plowed up a berm of sand and debris over a yard high, stripping acres of eelgrass, a shallows habitat, to bare sand in the process.

All these processes shape and are shaped by the estuary. They shape the bottom, the channels you use, the shoals you must avoid. Fifteen years ago, I became fascinated enough with this process that I made a cardboard model showing how a coastal inlet would look if dewatered. People were fascinated by it; it was simple to build, and, in fact, it hangs in my office as I write this chapter.

The process involved five steps, which I repeat here, and involved only pennies worth of materials. I'll wager the clever reader could do it in a long winter evening:

- I selected an area on the chart and 'contoured' it by connecting points of equal depth, interpolating where the little numbers were thin, to create a smooth curved line. NOAA usually does one contour six or twelve feet and shades in blue as an example for you. I chose vertical intervals of three feet so, that for this little bay, I ended up with six contours including the shoreline.

- If you use a last year's chart you can start in the center with the deepest contour and literally cut it out with a single-edge razor, or an Xacto® knife. Proceed with each of the other contours in the same manner.

- These pieces are then glued to shirt cardboard with a product like Elmer's® glue and pressed under a book overnight, with a sheet of wax paper to protect the book.

- Using a breadboard as a base, cut through the cardboard around the inside perimeter of each chart section, except the deepest ones, which should remain on a whole piece, equal in size to the chart section you're modeling. Discard the "hole" pieces, leaving a gap where water deeper than a given chart contour would be.

• Stack the contour pieces in their original order from top to bottom. Glue and press as before, and your model's done.

A former colleague liked my model so much he built one of Kettle Bottom shoals in the Potomac River; essentially the view from his house (without water). This is a nasty piece of riverbottom, and his model contained a dozen contours and scores of pieces. His wife, a computer programmer, wrote a routine that took all the numbers on their section of chart and plotted a 3-D picture of the bottom as it would appear from the air. That was ten years ago, and NOAA's charts will soon be using similar techniques to make charts that are an incredibly good representation of what the bottom below is really like.

We said earlier that the quantity of salt dissolved in water has important, sometimes life-threatening, implications for the estuary's living resources. The sea itself has an amazingly constant salt content "salinity," or, as it can be loosely defined, total dissolved solids. Along the eastern seaboard it probably won't vary 10 percent from an average 32 ‰ (parts per thousand).

To visualize this concentration, you can literally boil away a liter (1.06 quarts) of seawater and, in the bottom of the pan, find 32 grams (about 1⅐ ounces) of salts. In doing this you have removed 1,000 grams of water as steam; hence thirty-two "parts per thousand" of salt remain. Using the sun's heat to evaporate water in shallow ponds is how much of the world's commercial salt supply is generated. In colonial America, the process was a major "coastal cottage industry" from New England south. Saltworks were important targets in the American Revolution and the War of 1812.

All these salts mean that seawater is not only very different for aquatic creatures to live in, but that it's also substantially heavier than the freshwater flowing into the estuary from terrestrial sources. The fresher water literally floats on the saltier underneath, mixing relatively little at first. Of course, tides (chapter 14) slosh this system back and forth; wind and wave do their share of stirring and, on the general pattern of fresher-at-the-surface saltier-at-the-bottom, we find superimposed an overal gradient of increasing salt content as we near the ocean connection.

I track salinity weekly in the cove next to my boat using a very simple and inexpensive device, the hydrometer. It's almost identical to, but larger than, the float in those battery testers you see in gas stations. Hydrometers that read over the range of salts in seawater are available in many aquarium supply shops, or from many scientific supply houses for under eight dollars. The chart in this chapter allows you to convert the hydrometer reading to salinity if the water is at room temperature, and to detect changes in salinity as little as 1⁄10 part per 1,000.

You'll need a jar tall enough to float the hydrometer, 15–18 inches deep. I made mine by cementing a length of scrap plastic pipe to a square base using silicone tub caulk, and it's lasted seven years.

My salinity record shows a pattern typical for the average year, with a clear reduction in salinity accompanying snowmelt and the winter-spring rains, followed by an increase during the dry months until autumn. With coming rains, we see another decline. At this same location in the middle Chesapeake, I've seen salinity go from near fresh in tropical storm "Agnes" (1972) and "David" (1979) to over 19 ‰, about 60 percent seawater during dry summers when streamflow falls very low.

In the wet years we find freshwater insect larvae on the nearby sandbar, and fishermen can catch largemouth bass in the creek. In the dry years I can catch bluefish without ever leaving the mooring, and, at night, a rich saltwater plankton fills the water with sparkling luminescent dinoflagellates (plankton, chapter 11) and brilliant glowing comb-jellies (chapter 7).

Stratification, the layering of fresher waters over salt in more enclosed lagoons, like Barnegat Bay, and in deeper bays, like the Chesapeake, has made them especially good as storage systems for materials coming in from the surrounding watersheds. As materials settle out from the overlying fresh water, the incoming saltwater from below transports them back upstream to be deposited on the bottom. In colonial times, this ability to store and recycle nutrients meant a stock of food materials that helped make the waters productive. Now that

Conversion of hydrometer readings to parts per thousand salinity. Hydrometer, jar, and water must all be at 60°F (15°C) for the conversion to be accurate. Other extensive temperature-correction tables are available from the National Oceanographic and Atmospheric Agency, Washington, D.C.

(Density at 15°C.-- Salinity in parts per 1,000)

Density	Salinity	Density	Salinity	Density	Salinity	Density	Salinity	Density	Salinity	Density	Salinity
0.9991	0.0	1.0046	7.1	1.0101	14.2	1.0156	21.4	1.0211	28.6	1.0266	35.8
0.9992	0.0	1.0047	7.2	1.0102	14.4	1.0157	21.6	1.0212	28.8	1.0267	35.9
0.9993	0.2	1.0048	7.3	1.0103	14.5	1.0158	21.7	1.0213	28.9	1.0268	36.0
0.9994	0.3	1.0049	7.5	1.0104	14.6	1.0159	21.8	1.0214	29.0	1.0269	36.2
0.9995	0.4	1.0050	7.6	1.0105	14.8	1.0160	22.0	1.0215	29.1	1.0270	36.3
0.9996	0.6	1.0051	7.7	1.0106	14.9	1.0161	22.1	1.0216	29.3	1.0271	36.4
0.9997	0.7	1.0052	7.9	1.0107	15.0	1.0162	22.2	1.0217	29.4	1.0272	36.6
0.9998	0.8	1.0053	8.0	1.0108	15.2	1.0163	22.4	1.0218	29.5	1.0273	36.7
0.9999	0.9	1.0054	8.1	1.0109	15.3	1.0164	22.5	1.0219	29.7	1.0274	36.8
1.0000	1.1	1.0055	8.2	1.0110	15.4	1.0165	22.6	1.0220	29.8	1.0275	37.0
1.0001	1.2	1.0056	8.4	1.0111	15.6	1.0166	22.7	1.0221	29.9	1.0276	37.1
1.0002	1.3	1.0057	8.5	1.0112	15.7	1.0167	22.9	1.0222	30.1	1.0277	37.2
1.0003	1.5	1.0058	8.6	1.0113	15.8	1.0168	23.0	1.0223	30.2	1.0278	37.3
1.0004	1.6	1.0059	8.8	1.0114	16.0	1.0169	23.1	1.0224	30.3	1.0279	37.5
1.0005	1.7	1.0060	8.9	1.0115	16.1	1.0170	23.3	1.0225	30.4	1.0280	37.6
1.0006	1.9	1.0061	9.0	1.0116	16.2	1.0171	23.4	1.0226	30.6	1.0281	37.7
1.0007	2.0	1.0062	9.2	1.0117	16.3	1.0172	23.5	1.0227	30.7	1.0282	37.9
1.0008	2.1	1.0063	9.3	1.0118	16.5	1.0173	23.7	1.0228	30.8	1.0283	38.0
1.0009	2.2	1.0064	9.4	1.0119	16.6	1.0174	23.8	1.0229	31.0	1.0284	38.1
1.0010	2.4	1.0065	9.6	1.0120	16.7	1.0175	23.9	1.0230	31.1	1.0285	38.2
1.0011	2.5	1.0066	9.7	1.0121	16.9	1.0176	24.1	1.0231	31.2	1.0286	38.4
1.0012	2.6	1.0067	9.8	1.0122	17.0	1.0177	24.2	1.0232	31.4	1.0287	38.5
1.0013	2.8	1.0068	9.9	1.0123	17.1	1.0178	24.3	1.0233	31.5	1.0288	38.6
1.0014	2.9	1.0069	10.1	1.0124	17.3	1.0179	24.4	1.0234	31.6	1.0289	38.8
1.0015	3.0	1.0070	10.2	1.0125	17.4	1.0180	24.6	1.0235	31.8	1.0290	38.9
1.0016	3.2	1.0071	10.3	1.0126	17.5	1.0181	24.7	1.0236	31.9	1.0291	39.0
1.0017	3.3	1.0072	10.5	1.0127	17.7	1.0182	24.8	1.0237	32.0	1.0292	39.2
1.0018	3.4	1.0073	10.6	1.0128	17.8	1.0183	25.0	1.0238	32.1	1.0293	39.3
1.0019	3.5	1.0074	10.7	1.0129	17.9	1.0184	25.1	1.0239	32.3	1.0294	39.4
1.0020	3.7	1.0075	10.8	1.0130	18.0	1.0185	25.2	1.0240	32.4	1.0295	39.6
1.0021	3.8	1.0076	11.0	1.0131	18.2	1.0186	25.4	1.0241	32.5	1.0296	39.7
1.0022	3.9	1.0077	11.1	1.0132	18.3	1.0187	25.5	1.0242	32.7	1.0297	39.8
1.0023	4.1	1.0078	11.2	1.0133	18.4	1.0188	25.6	1.0243	32.8	1.0298	39.9
1.0024	4.2	1.0079	11.4	1.0134	18.6	1.0189	25.8	1.0244	32.9	1.0299	40.1
1.0025	4.3	1.0080	11.5	1.0135	18.7	1.0190	25.9	1.0245	33.1	1.0300	40.2
1.0026	4.5	1.0081	11.6	1.0136	18.8	1.0191	26.0	1.0246	33.2	1.0301	40.3
1.0027	4.6	1.0082	11.8	1.0137	19.0	1.0192	26.1	1.0247	33.3	1.0302	40.4
1.0028	4.7	1.0083	11.9	1.0138	19.1	1.0193	26.3	1.0248	33.5	1.0303	40.6
1.0029	4.8	1.0084	12.0	1.0139	19.2	1.0194	26.4	1.0249	33.6	1.0304	40.7
1.0030	5.0	1.0085	12.2	1.0140	19.3	1.0195	26.5	1.0250	33.7	1.0305	40.8
1.0031	5.1	1.0086	12.3	1.0141	19.5	1.0196	26.7	1.0251	33.8	1.0306	41.0
1.0032	5.2	1.0087	12.4	1.0142	19.6	1.0197	26.8	1.0252	34.0	1.0307	41.1
1.0033	5.4	1.0088	12.6	1.0143	19.7	1.0198	26.9	1.0253	34.1	1.0308	41.2
1.0034	5.5	1.0089	12.7	1.0144	19.9	1.0199	27.1	1.0254	34.2	1.0309	41.4
1.0035	5.6	1.0090	12.8	1.0145	20.0	1.0200	27.2	1.0255	34.4	1.0310	41.5
1.0036	5.8	1.0091	12.9	1.0146	20.1	1.0201	27.3	1.0256	34.5	1.0311	41.6
1.0037	5.9	1.0092	13.1	1.0147	20.3	1.0202	27.5	1.0257	34.6	1.0312	41.7
1.0038	6.0	1.0093	13.2	1.0148	20.4	1.0203	27.6	1.0258	34.8	1.0313	41.9
1.0039	6.2	1.0094	13.3	1.0149	20.5	1.0204	27.7	1.0259	34.9	1.0314	42.0
1.0040	6.3	1.0095	13.5	1.0150	20.6	1.0205	27.8	1.0260	35.0	1.0315	42.1
1.0041	6.4	1.0096	13.6	1.0151	20.8	1.0206	28.0	1.0261	35.1	1.0316	42.3
1.0042	6.6	1.0097	13.7	1.0152	20.9	1.0207	28.1	1.0262	35.3	1.0317	42.4
1.0043	6.7	1.0098	13.9	1.0153	21.0	1.0208	28.2	1.0263	35.4	1.0318	42.5
1.0044	6.8	1.0099	14.0	1.0154	21.2	1.0209	28.4	1.0264	35.5	1.0319	42.7
1.0045	5.9	1.0100	14.1	1.0155	21.3	1.0210	28.5	1.0265	35.7	1.0320	42.8

man has occupied so much of all our watersheds, it also means that pollutants and excess enrichment also tend to accumulate, often with catastrophic results.

Back in the 1950s duck-farming operations contaminated the drainage of Moriches Bay on Long Island's south shore. Tidal flushing through a small inlet to the sea was very slow, and nutrients accumulated, fertilizing a population explosion of tiny algae cells which proved unpalatable to the bay's commercially valuable oysters. They literally starved to death in an indigestible soup!

Both agriculture and urbanization have increased inputs of nitrogen and phosphorus (and of toxics like lead from automobile gas) that enter Chesapeake Bay from her tributary rivers. The result has been increasing plankton growth over the years. Cells that die settle to the bottom and begin to decompose there in the deep central channels where tidal mixing normally doesn't reach them. This decomposition—rotting—consumes tremendous amounts of oxygen from the water, literally killing bottom organisms and driving fish into shallower water to escape suffocation. Since the 1950s the portion of the bay that became 'anoxic' (devoid of all oxygen) each summer has increased something like ninefold. With man and all his works, including sewage-treatment works, in place, the solutions to the bay's serious problems are not easily found. The Chesapeake experience with lowered bottom oxygen is present in many other bays, though not usually to such a remarkable extent.

My first experience with a nutrient-mediated plankton bloom was also my introduction to marine biology. During the summer and fall of 1964, huge patches of dinoflagellate "red tide" covered much of Barnegat Bay in New Jersey, and as the semimicroscopic plant cells died, the process of decomposition rapidly depleted oxygen from the shallows, driving fish and crabs, desperate for a clean breath, right to the water's edge. Clams, trapped on the bottom, extended their siphons and died. Rotting masses of plankton, fish, and other critters released enough odiferous hydrogen sulfide to discolor the paint on houses near the shore.

In many estuaries, from Boston Harbor to Florida's Biscayne Bay, we're realizing, maybe too late, that our society's impact on natural systems flies right in the face of our self-interest in having an estuary that meets our aesthetic, recreational, and culinary needs. These are, after all, the things that draw us all to water in the first place; why then use our waterways as a repository for wastes? Estuaries are, and really must remain, water courses for most of our nation's commerce: from Datsuns to dynamite, Kuwait crude to uranium ore. Our greatest cities are on estuaries for just that reason and have served as lodestones for population over the last 375 years.

We're all a part of the problem that estuaries face. We all create garbage, flush toilets, use electricity, and burn oil. It will challenge our own and future generations to plan our consumption and moderate our impacts so that estuaries will not just hold the line, but can be put on a course toward restoring the productivity they have had for past generations.

SUGGESTED READING

The World of an Estuary, by Heather Angel (London: Faber & Faber, 1974).

Ecology of Inland Waters and Estuaries, by George K. Reid (New York: Van Nostrand/Reinhold, 2nd edition, 1976).

Kent Mountford, born in 1938 in Plainfield, New Jersey, earned his Ph.D. in 1971 from Rutgers University. His research interests, then and now, stem from his "abiding love of the plankton, that great system of myriad forms that lies as interceptor between our energy source, the sun, and the ecosystem earth." He continues as to his purpose: "It is for the phytoplankton that, self-appointed, I've become interpreter: to impress upon industrialized unfeeling man the debt of understanding and fidelity he owes these uncounted billions he cannot see."

Dr. Mountford is a naturalist specializing in estuarine watersheds of the Chesapeake, a biological photographer and illustrator, a consultant to environmental-impact projects, a boater and teacher, and a lecturer and author.

Although a lobster's natural color is a dark greenish-brown, there are dozens of freaks of nature. NEW YORK ZOOLOGICAL SOCIETY

Barnacles feeding NEW YORK AQUARIUM

Colony of brown sea anemone NEW YORK AQUARIUM

Powder puff sea anemone NEW YORK AQUARIUM

Periwinkles NEW YORK AQUARIUM

The Spanish dancer nudibranch is an example of the most delicate and lovely of the mollusks.
NEW YORK AQUARIUM

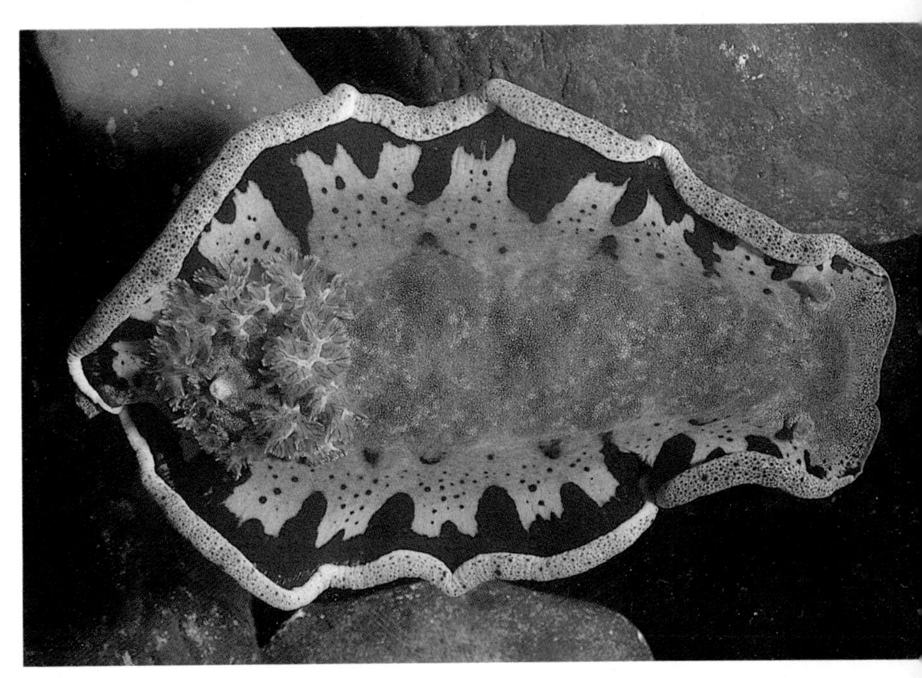

The most common sea star on the Atlantic coast is Forbes' asterias, *found from Cape Cod to the Gulf of Mexico.* NEW YORK AQUARIUM

Thorny oysters grow on rock ledges and shipwrecks off eastern Florida at depths of five to fifteen fathoms. AMERICAN MALACOLOGISTS, INC.

The fragile purple sea snails are common pelagic voyagers in the Gulf Stream. Self-made rafts of bubbles keep them afloat until they reach some distant beach.

AMERICAN MALACOLOGISTS, INC.

After boiling shells for six minutes, twist out the meat. Freezing overnight will also do the trick.

R. TUCKER ABBOTT

Flowering tropical turtle grass SYLVIA EARLE

Halimeda, *a calcareous green alga, contributes a major portion of the limy matrix to coral reef formations.* SYLVIA EARLE

Some species of seaweed have bizarre shapes, such as the warm-water green grape look-alike Caulerpa racemosa *and mermaid's wineglass.*
SYLVIA EARLE

The common kelp washed up on your beach is actually an extraordinary plant. Some strands grow as much as a foot a day and attain a length in excess of 150 feet. SYLVIA EARLE

Estuaries are semi-enclosed bodies of water that retain free interchange with the sea and are partly diluted by freshwater. Some estuaries, like those in Fire Island, New York, are more distinctly enclosed by wave-built islands or peninsulas. KENT MOUNTFORD

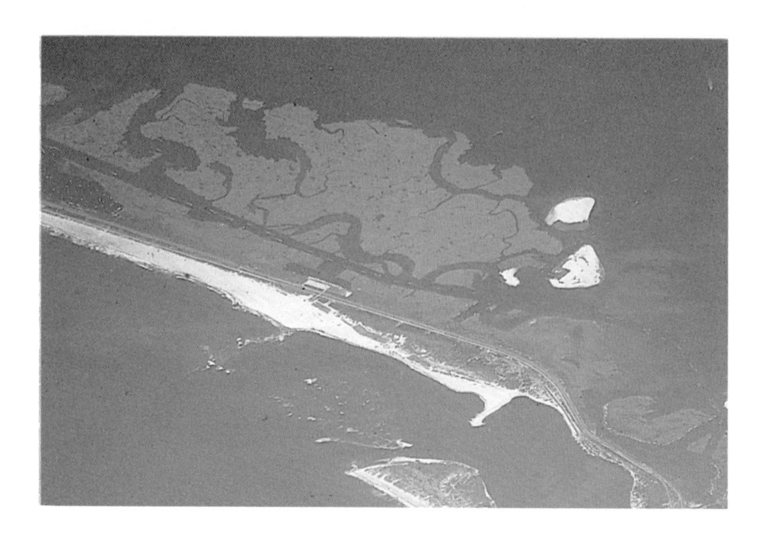

Pike fish DR. STEPHEN SPOTTE

Red starfish DR. STEPHEN SPOTTE

Flame scallop NEW YORK AQUARIUM

Chambered nautilus NEW YORK AQUARIUM

The LAND BELOW and BESIDE the SEA: COASTAL GEOLOGY

Dr. Robert F. Dill

□

All persons wishing to understand *coastal geology*, be they yachtsman, fisherman, engineer, or scientist, will find it helpful to be able to describe the physical nature of the ocean floor with a common vocabulary. Definitions are therefore a necessary part of all communication when working with the sea. Marine geologists have developed specific terms and use photographs and maps to describe features found in the offshore region and in discussing the continental margin. I hope that as you read this chapter on coastal geology that you will acquire a better feeling for that world those of us who work there find so fascinating.

Marine scientists, when contouring the first crude offshore bathymetric charts—measurements of ocean depths, made by plotting individual lead-line soundings—or later, when developing modern charts constructed from precisely located bottom profiles made from continuous echo soundings across the continental margin, discovered that this large region could conveniently be separated into areas of similar slope, shape, and depth. This division has aided understanding the location and nature of marine processes, types of sediments, and the ecological zonation of organisms found in this region of such great importance to us all. The marine margin off the East Coast shows considerable variation between bottom profiles made off Maine and those made off Florida. However, all profiles from the shore out to deep water have a bottom configuration with features common to all. Starting at the shoreline and moving seaward, all depth measurements show a gently sloping plain separating the coast from the deep ocean. This common feature is called the Continental Shelf. It is not uniformly of the same width or shape, nor is its outer edge defined by a specific depth contour before falling into the deep ocean. To further complicate matters, besides bearing physical dissimilarity from one region to another, the boundaries of this shelf have also been defined differently by scientists than by legislators or governmental offshore regulations, enacted to specify boundaries of pollution control and claimed offshore resources (e.g., fish, oil, gas, etc.). Legal definitions of the Continental Shelf were originally drafted by attorneys to describe areas covered by laws to lease offshore oil and gas deposits claimed by the U.S. government. Many of these lawyers were frequently unaware or unmindful of the physical differences existing from one region to another. Consequently, legal boundary definitions often remain unrelated to the real world described by scientists. Thus, when a yachtsman talks with or reads articles published by the marine scientist describing events observed or features discovered on parts of the continental margin, such as the Continental Shelf, this may or may not coincide with the area described by the attorneys or bureaucrats, who write the laws claiming legal authority over specific activities in an offshore region.

To clarify scientific terminology used in this chapter on coastal geology for the reading layman, I present the following definitions extracted from the glossary of geological terms initially compiled by leading marine scientists for the American Geological Institute, with subsequent revision in 1981.

> *The Continental Shelf:* "that part of the *continental margin* extending seaward from the low-water line to a depth (usually between 100 and 200 meters) at which there is a marked increase of slope or a rather steep descent toward greater oceanic depths. It is characterized by its overall gentle slope of less than 0.1°."

> *The Outer Continental Shelf Break:* "the seaward edge of the Continental Shelf where there is a zone of slope transition from gentle to steep."

The Continental Slope: "that part of the continental margin that is between the *Continental Shelf* and the *Continental Rise*, or *Oceanic Trench* or *Abyssal Plain*. It is characterized by its relatively steep slope of between 3–10°."

The Continental Rise: "the upper surface of a wedge of sediment that extends from the base of the continental slope out to the deep-sea floor (it is characterized by a more gentle slope than the Continental Slope)."

The Abyssal Plain: "the very flat regions of the deep-sea floor, usually at the base of the Continental Rise. It is characterized by ponded sediments carried to the deep ocean through submarine canyons by gravity-driven currents."

Another feature seen on bottom charts that requires some mention is the *Blake Plateau*. It lies seaward of the *Shelf Break* at depths between 500 to 1,000 meters as a broad, irregularly surfaced terrace that probably owed its origin to the northward flow of the Gulf Stream. The Blake Plateau starts south of Cape Fear at about Latitude 33° and extends about 400 nautical miles in a southerly direction down to the Bahama Island Platform.

For the mariner to understand coastal geology, seen as he looks shoreward from the deck of his boat or reflects about what is shown on his charts, requires an additional knowledge of what has happened to the margins of the continents since the last Ice Age. Most of the features seen on navigational charts, the shape of the harbors, etc., including the movement of coastal water masses, have been affected in some manner by what occurred in that frigid period of the past. Starting approximately 30,000 years ago, the earth's surface began to cool. Weather patterns changed and vast quantities of seawater were drawn from the world's oceans to form thick ice sheets. These vast plains of ice (up to two miles thick) covered much of the continental land masses both in the Northern and Southern hemispheres. The withdrawal of water from the ocean basins and buildup of ice reached its maximum about 18,000 years ago, lowering sea level far below its present stand, exposing much of what is now the submerged Continental Shelf to the erosion of rivers, rain, and, in the higher latitudes, the passage of glaciers. Extensive forests, open marshes, and coastal plains with dunes and deltas existed where today we sail and anchor our boats. Many of the bottom features depicted on navigational charts were originally land or shoreline forms, relicts of this former time of sea-floor exposure. Similarly, many of the features that form the coastlines of today reflect this time of lowered sea level, representing modified waterways, riverbeds, ancient deltas, and erosional remnants of the glacial period recently passed. In the eyes of the geologist who deals with an earth history of hundreds of millions of years, the exposure of the offshore continental margin 18,000 years ago is but a moment of yesterday. However, for the sailor, it was a most important moment, because it was a time of formation or change for much of what we see on the present coast and measure with our instruments in the shallow seas.

On the east coast from Long Island, New York, north to beyond the Canadian border, glacial debris gouged off the hard rock surfaces of the continents by ice sheets was carried out onto and in some regions to the very outer edge of what is now the Continental Shelf. Areas such as Georges Bank are blanketed by glacial material, deposited as far as 150 miles seaward of the present coastline. Now, in the shoaled waters, tidal currents that rotate in an elipse every twelve hours carry this relict pile of sand and gravel first north, then south, grinding rock grains smooth and piling them up in ever-changing submarine dunes. Closer to shore at Nantucket Shoals, the interdunal areas channelize these tidal currents and cause strong rips and dangerous water over the crescent-shaped shoals made by the large boulders and sands dropped and left behind from the gouged-up material bulldozed to the sea by the glaciers. During maximum glaciation, sea level was lowered to a depth that placed the shoreline with its sand beaches near the *Shelf Break*, the outermost edge of the shelf. Rip currents and strong longshore currents diverted land-derived sediments seaward beyond the surf zone. Where this glacial material was rapidly dumped over the upper edge of the outer shelf onto the steeper Continental Slope, it became unstable. Giant slumps and sediment-filled marine avalanches eroded the continental margin. Sediment-charged bottom currents flowing downslope eroded giant submarine canyons all the way down to the very deep water at the base of the continent. There the slope decreases and the sediments fall out of suspension as currents flow with ever-decreasing velocity out and onto the gently sloping deep-sea floor forming the Continental Rise and deep-sea fans. The extensive submerged canyon system that diverted the shallow water sediments into deep water extends all along the entire outer margin of the east coast as far south as Cape Hatteras. The huge amount of sediment derived in shallow water and from the erosion of canyons filled in deep-sea basins and muted the topographic highs to form the flat "abyssal plains" found at the base of the continental margin in depths down to 15,000–18,000 feet and as far as 600 miles out into the Atlantic Basin.

Looking at a detailed bottom chart of the outer margin of the east coast (for example, National Ocean survey Bathymetric Map 13003), you will discover that the heads of many of the large submarine canyons hook northward. This is caused by canyon heads being eroded by sediment spilling into them and moving downslope by submarine processes such as slumps, slides, bottom currents, sand flows, etc. The submarine canyons cease to be eroded shoreward when they reach the surf zone because the heads fill with the sands moving along the coast. The canyon head will then begin to hook toward the source of sediment, because sand spilling over and eroding the upper lip of the canyon head is more active on the side of the canyon head receiving sediment than on the down-current side. Thus, the shape of the east coast canyons tells us something very significant about the position of sea level during the last Ice Age and the direction of sand movement along ancient beaches. Further, canyon shape reveals that sea level must have been at least 330 feet below its present level to provide the large volume of moving beach sands needed to erode the hook in the heads of the canyons.

Further examination of the offshore bathymetric chart will show that the Continental Shelf 18,000 years ago was much narrower than at present. It was then much like the west coast of the United States off southern California or the southeast coast of Florida off Miami, where now a short sail of less than ten miles can reach deep water. Now, to the contrary, on the east coast from Cape Canaveral north, a sail of fifty to one hundred miles is required to reach comparable depths because of the gentle seaward slope of the broad shelf. The canyon heads are relatively inactive, receiving no sediment from the present beach. It is interesting to note that submarine canyons off southern California are cut to the surf zone and are actively diverting beach sand into the deep sea.

A short time ago, "geologically speaking," areas now covered by water were broad coastal plains, grazing areas for the "woolly mammoth," giant mastodons, tapirs, musk ox, giant moose, wild horses, and ground sloths. Early man, when he reached the New World, probably roamed this vast forested plain and hunted these now-extinct species. Some speculate that he may have been responsible for their demise, but a rising sea level and ever-shrinking habitat must also be considered when conjecturing. Fossil bones and shallow-water shells representing these Ice Age animals, along with wood remnants, bog peat, and pollen-rich mud from the forests now drowned, are occasionally dredged up in fishing trawls and in the sampling gear scientists use when exploring the offshore areas of the east and gulf coasts.

After the great chill that created the last Ice Age, and starting about 15,000 years ago, the world's climate began to warm. The vast amounts of water released by the melting and ever-retreating glaciers flooded the riverways and carried the sediments gouged up by the glaciers all the way to the shoreline. Because sea level was rising as water locked up on land returned to the ocean, the shoreline rapidly advanced shoreward across the coastal plain, flooding land features with seawater. Large quantities of sand, boulders the size of houses, and mud were rapidly brought to this advancing shore to be reworked by giant waves generated during severe storms. The fine-grained muds, soil, and glacial powder that formed a large fraction of this runoff was deposited as a blanket of fine clay in depressions on the shelf and in the deeper quiet waters beyond the surf zone. With time, this fine-grained material formed a stiff clay that entombed the shell bodies of the cold-water clams, oysters, and other burrowing shellfish, forming a distinctive fossil record that can be recognized in cores and dredge samples as dissimilar from warmer-water forms now living in today's sediment. The core or dredge described in the chapter of this book entitled "Do-It-Yourself Oceanographic Equipment" can be used to find your own fossils from this relict deposit. Examine low areas between offshore ridges for your samples. The fossil shells frequently have a grayish-to-black-colored powdery surface or a dark brown luster from a phosphorite ocean-mineral coating.

The sands and coarser-grained components of the runoff became barrier beaches that grew in size until they became large elongated islands and peninsulas. Excess sands pushed up beyond the wave zone have been blown shoreward by coastal winds, then stabilized by plants, and now form the dunes of today's shoreline that extend all the way south to the Florida Keys. Gravels and boulder-sized sediments, too heavy to be carried by coastal currents, were deposited at rivermouths to form the headlands of barrier banks and islands such as Cape Hatteras, Cape Fear, and Cape Lookout.

To better visualize the effects of the change of sea level, imagine you are standing at the present shore 18,000 years ago viewing a "time lapse," a speeded-up version of time, and could view the period following the last Ice Age in less than the daylight hours of a summer's day. If your timer started at 0600, to the east you would see forests and broad meadows forming a vast plain. You would be standing approximately 300 feet above sea level. The beach would be far to the east; for instance, off New York, it would be about 80 miles away. The Hudson River would be winding across this broad coastal plain, and a dense forest would stretch beyond sight. To the north, a cliff of ice forming the edge of a large continental ice sheet would cover Long Island all the way to the river's edge. At about 0800 (15,000 years ago), the day would warm up, and water released from the ice sheets during periods of fast melting would cause sea level to rise rapidly and flood westward across the gentle slope of the coastal plain. A small rise in sea level would result in a great distance of westward displacement. At about 1100 (11,000 years ago), the shore will have moved about 20 miles west and you would be only 210 feet above sea level. During periods of slow melting, sea level would remain in place, forming long beaches and dune systems, and where topographic relief was high, sea cliffs would form. These features would remain behind as flooded bodies and relicts of a former of stand of sea level when sea level again rose during rapid melting as the weather again warmed. We know from observing samples of such features on today's sea floor and from radioactive carbon dates of fossil wood and shells found in them that the release of water from the ice was not a gradual event, but one of fast melt followed by a period of stability.

As your speeded-up day progressed you would see the level of the sea rise in a series of steps, across the forested plains, flooding and leaving intact the long linear beach lines and low cliffs that are presently recognizable as north-south trending ridges on bathymetric charts of the sea floor. At about 1600 (4:00 P.M.) you will be standing near the shoreline, and sea level will not change until sunset, as it stabilized after the melting of the major ice sheets. The features just described can be seen on your echo-sounding tapes if you run east-west across the shelf. Samples of the ridge crests and of troughs between should bring back shells of shallow-water organisms that do not presently live at these depths. These are the fossils of the Ice Age mentioned earlier. The oldest are deepest and furthest from the present shoreline. Thus, many of the shoals and irregularities seen on the navigation charts represent these flooded coastal features left behind by this "transgression" of the seas, a term used by oceanographers to describe the shoreward movement of a flooding sea.

THE DYNAMIC BEACH AND CHANGING COASTLINE

Much to man's chagrin, the beach that he loves to "sun on" and "swim from" or "build next to" is not stable, but an ever-moving and changing dynamic body. Its shape and form depend on a continual supply of new material to replace the ever-moving sand grains that constitute its seaward face. Never stable, the sands are washed first up and then down the face by waves striking the beach and are moved from one point to another by prevailing longshore current. When you next visit the beach, take time to watch the movement of this sand. Note the "swash marks," small, linear ridges of depris and sediment that mark the uppermost shoreward flow of water after each wave. Watch how the water level sinks below the surface of the highly permeable beach when the sea level retreats to form the trough of the next wave. If the beach becomes saturated, the trapped water will flow from the sands. "Rill marks," V-shaped erosion scars, will form as the water contained in this submerged water table flows out and down the beach. The sands are also moved by water currents generated by storm swell, wind waves, and the tidal rise and fall of the sea. The sand areas exposed above water are often blown shoreward by the winds to form dunes. The shoreline of the east coast lying in the path of hurricanes is also subject to catastrophic changes in sea level. The rapid decrease in atmospheric pressure, represented by a falling barometer, will cause a seawater level to rise up to ten and even fifteen feet above normal levels. This bulge of the sea level surface is called a "storm surge" and is a response of the sea's surface upward to the extremely low air pressures in the "eye of a hurricane." If a hurricane hits the coast during high tide, massive flooding of low coastal areas and lagoons, accompanied by high wind-driven waves, occurs. These trapped waters generate strong currents as they flow through narrow tidal inlets when the water surface again returns to normal levels. Massive movement of the sands forming the beaches and offshore bars follows. The coast, with its offshore islands and tidal inlets, is often reshaped after the passage of a hurricane. These changes are so severe and the following readjustment of the coastline so altered when normal conditions return that the National Ocean Survey, which is the U.S. government's coastal mapping agency, will not attempt mapping and recharting efforts for over two months following a major hurricane. Thus, errors in chart depths and shapes of coastal inlets can be possible and must be anticipated by the boater following severe storms, particularly hurricanes.

It is easy to visualize why beaches move and shorelines change during the catastrophic development of waves and currents with the passage of a hurricane. But we also observe that the shore changes in a more subtle way over a daily and yearly cycle of time when its seaward face is subjected to normal tidal and wave periods. An entire body of scientific knowledge has grown up around studies made to determine the causes of coastal change as beaches build in one area and erode in another. Scientists and engineers conducting research on beaches are funded by the U.S. Army Corps of Engineers, the National Science Foundation, the National Oceanic and Atmospheric Administration, and the U.S. Navy. Funds from these agencies are used by university and government scientists all along the coast of the eastern United States to study the effects of waves, tides, wind, and coastal vegetation on sediment movement. Sediments entering the coastal region are monitored and traced to provide the yachtsman with the latest information and conditions that will affect his boating ability. Published as scientific reports, this information is available to the boater and can be invaluable in developing an awareness of coastal geology, the processes active in this region, and insight into why changes take place along the coastal zone. A list of recommended articles and books giving general coverage of interest to the mariner is provided at the end of this chapter.

Model studies with laboratory-created waves of given heights and shapes striking shores of different configurations and composition are made to develop a knowledge of what is known as "coastal dynamics." These data are used to predict what will happen when man tampers with "natural" movements by creating breakwaters, dredging channels, and enlarging lagoonal and modifying lowland regions. Unfortunately, the history of man's use and resulting modification of the dynamic system that is the shoreline has not been one that he can be proud of. It has not been until the past two decades that the knowledge and ability to understand coastal processes have grown to a point where intelligent and rational results have been achieved. Even now, with the scientific sophistication provided by advancing computer technology, satellite imaging, new instrumentation, and a long recorded history of change in shoreline features and the elevation of sea level, we are only approaching an understanding of beach and shoreline dynamics.

It is most important to understand the dynamic nature of the shoreline before investing in a region, either in berthing for your boat or in a coastal residence. It is wise precautionary insurance to seek out the problems that could affect your investment by asking an appropriate scientist or engineer for an opinion. Check your nearest coastal university—most have professional scientists on their staff who will feel privileged to venture an opinion that could save you money in the future. It is also prudent for the passing yachtsman to ask local fishermen who spend their time working on the sea, observing the day-in and day-out changes, what conditions currently prevail. Be wary—it's your boat and responsibility, so don't rely on everything published. Astute scientists who work in coastal regions often inquire and are continually guided by local "watermen" about the conditions and changes that occur in an area of study. Many of today's recognized oceanographers preceded their professional careers as fishermen or surfers. During their youth they learned the feel of the sea that they now describe in complex mathematical formulas.

When one thinks of eastern coastal geology, the dominant region of concern is usually the beach or the extensive north-south waterway running behind it. Most often the beaches are formed of sand-sized grains (.062 to 2 mm in diameter), but may be cobble or pebble size (above 2 mm) in areas of high energy and where there is a source of this coarse-sized material that cannot be carried away by longshore and wave-generated currents. Often the source of coarse-grained beach material is from ancient deposits exposed in low sea cliffs behind the beach. In geology, recycling is common, with sands, cobbles, and gravels often going through many periods of erosion as they are ground down into finer and finer particles with each passage through the surf zone. The beach is defined as: "that area extending from the low-tide waterline to the crest of a dune or to the base of a cliff, if one is present, or to the seaward side of the line of permanent vegetation." The region of the beach is often divided by the coastal geologist into two zones: the foreshore and the backshore, depending on its exposure to tidal and wave-induced sea level change. In general, the dynamics of beach-sand movements and the shape or steepness of the beach foreshore depend on the height and direction of wave approach, tidal range, and the grain size of the sediment forming the beach. A rule of thumb is that: "coarse sand beaches have steep forebeach slopes; fine-grained sands form gentle beach slopes."

Waves breaking at the shore continually push the coarser grains of sand and cobbles landward; however, if they strike the beach front at an angle, they generate a longshore current in the surf zone that moves the beach sediments "down coast." The currents set up in the surf zone where the waves strike the beach at an angle are known as "littoral currents." Along the east coast, the dominant flow is to the south, because the waves most responsible for beach movement come from the northeast during large winter storms ("northeasters"). However, flow can be reversed in local areas with changes of shoreline orientation and an increased dominance of tidal currents over wind-driven currents. Another manifestation of the southward drift of sediment can be seen in the nature of the sands forming the beaches. The yachtsman sailing south will discover that the sands change in composition, becoming more quartz-rich as they are transported away from local rivers which are areas of influx of sediment to the beach. The northern glacial source is also diminished by intergranular erosion as it moves south.

Southern movement is also evident on the northern side of "groins,"* frequently employed to trap sands to shore up the shoreline in front of the residential beach properties. This, of course, cuts off the supply of sands to the properties downcurrent, with adverse effects. Wave action will continue to move the existing sands away from the down-current beach and it becomes "starved" of sand: erosion begins shoreward; shore properties with their expensive dwellings are no longer protected by a broad beach and are destroyed by winter waves;

*A *groin* is a man-made structure, quite often a concrete wall or wooden bulwark extending seaward perpendicular to the beach and designed to slow, trap, or dam the longshore movement of sand in the surf zone.

lawsuits develop; political pressure is exerted, and a very costly beach replenishment project may result—paid for in many instances by public funds. Unfortunately, this cycle has been a recurring event along most coastal regions over the past century—a period when man developed a desire for the pleasures of beach living, but until recently gave little thought to the dynamic processes that govern beach behavior.

Small-boat harbors that have been built behind the barrier islands and in the lagoons are other areas where sand movement is also critical. The well-being of a small-boat harbor depends on maintaining an open channel to the ocean. The moving sands have a tendency to fill these channels; consequently, breakwaters are constructed on the up-current side to trap moving sands and down-current to constrict the tidal-induced flow of water in and out of the lagoon behind the barrier islands. The design of breakwaters to stop channel migrations, along with the computations of the "tidal prism," the volume of seawater flowing in and out of the lagoons, is complex. If executed correctly, the water flowing in and out of the lagoon keeps channels open and brings about long-term stability All factors of sediment dynamics are used in these designs. Over the past decade, scientists and engineers responsible for coastal studies have begun to show positive evidence of understanding the processes involved in today's coastal development projects. Although not perfect, when compared with many of the projects of the past, the more recent programs of beach development at least reflect an awareness of past problems and give promise to improve future utilization of the coastal zone as man's influences increase.

Moving farther south, the beach is separated from coastal highlands by the extensive swamps and estuarine systems of South Carolina and Georgia. Here, rivers flow slowly, causing them to dump much of their sedimentary loads in the swamps before reaching the shoreline. On the coast, this initiates the beginning of a change in the nature of beaches. Also, different types of sediment begin to dominate the beach. By mid-Florida, broken shells and coral debris washed from offshore now become the important constituents, because the quartz grains, no longer added from former river sources, begin to be eroded away. With the warmer waters of the lower latitudes and occasional influxes of Gulf Stream eddies, the productivity of shell-bearing and reef-building organisms increases, along with a decrease in the solubility of the calcium carbonate ($CaCO_3$), the main constituent of marine shells, corals, and even plants living in seawater. At approximately Key Biscayne, just south of Miami Beach, the quartz sands derived from the rivers and glacial sources of the north are almost extinct. From Key Biscayne south to Key West, except for a few small patches on restricted beaches, the sediment found both on the beach and offshore owes its origin to marine plants and animals, not to the rivers of the north.

The beaches of the southern coast of Florida are washed by warm seawater that is saturated with calcium carbonate. These waters, when trapped between the grains of sand forming the beach, are involved in a complex chemical interchange that has altered the nature of some beaches. If conditions are precisely proper—a change of temperature, the amount of seawater mixed with organic-rich rainwater, or a period of evaporation of damp sand above sea level—a hard chemical precipitate will form that cements the grains of sediment together, forming a hard rock called "beach rock." This cemented beach forms rapidly and contains the fossils, even old bottles and man-made "junk" that were in the sand initially forming the beach. Even the bedded nature of grains forming the original slope of the beach face is preserved. Sands not cemented are carried away by current; hard ledges of tabular rock remain as exposed linear pavements sloping seaward between low- and high-tide water levels. The shoreline prominences and headlands of the Keys and, across the Straits of Florida, in the Bahama Islands are often formed of resistant "beach rock." Although in places still forming, much of the resistant beach rock seen on today's beaches is much older. It formed 2,000 to 4,000 years ago, when the sea surface rose rapidly to about—but slightly below—its present level.

The relatively stable position of sea level at nearly its present position over the past 4,000 years is not unique for the 18,000-year period following the last Ice Age. However, equivalent "still stands" of sea level were not of such a long duration. Even so, these "still stands" did leave their mark on the sea floor, because, in the past as now, beach rock formed in tropical waters, creating long, linear beachrock-defended shorelines now flooded. The broken remnants form hiding places and have become the domain of tropical fish. These cemented rocky areas, preserved as terraces that can be observed in shallow waters by snorkeling and in deeper waters by scuba diving, are known as "deep reefs"—favorite fishing areas off Florida.

There is a place near Bimini Island in the Bahamas where these submerged beaches have been misinterpreted as relicts of ancient highways leading to the fabled "City Civilization of Atlantis." Carbon 14 dates of shells in the "pavement" conform to the time of the existence of Atlantis. The features are long and linearlike highways, and, most important, people who don't know the true origin or fail to look at today's beaches want to believe in the existence of an ancient civilization off Florida—it's good for tourism even if not true!

The hard rock substrates formed by the "beach rock" in the past, but now in deep water because of sea level rise, provide a stable anchoring ground for coral growth. Important ancient sea level stands are found at depths of approximately twenty, forty, and sixty feet below the present level. After sea level rose to its present location, the coral and the complex organic community that moved in and colonized these rocky areas built mounds of cemented skeletal material on the sea floor. In some areas, the coral species forming the innermost submerged reef have now reached the present sea level. The results have been a series of linear offshore reefs that run the length of Florida and are called "bank barrier reefs." These reefs act as an offshore breakwater, separating the high waves of the open sea from today's shoreline by a broad shallow lagoon filled with organic sands, marine grass beds, and "patch reefs." The latter formed on a rock base before the offshore bank barrier reef reached the surface. Boaters often use the quiet waters of the lagoon as highways of safe passage when sailing down the Keys from south Miami. Thus,, when looking at a chart of any tropical region, be it Florida, the Bahamas, or the islands of the Caribbean, the yachtsman can recognize and understand how the worldwide fluctuating sea level of the last Ice Age dramatically formed the present coastline, especially in tropical regions.

COASTAL FORMS NORTH TO SOUTH

Recognizing that the coastal region has been modified by sea level changes over the past 30,000 years and that the form of the coast, its "geomorphology," will also reflect the rock type or geology of the coast, one can begin to comprehend the great changes in shape, elevation, and composition of beaches that are found as one sails along the east coasts of the United States, from the "rocky shores" of Maine to the low islands of the central coast and the long coral reefs, lagoons, and beaches of the Florida Keys.

The New England coast from Maine down to New York is cut into an ancient hard rock region hundreds of millions of years old that has suffered through geological time by being crumpled into mountain chains. These were later eroded irregularly by rivers to form valleys and low hills—the sediment from this erosion formed a broad coastal plain. During the Ice Age, the rocky areas were further ground down and smoothed by the passage of thick continental ice sheets. Calculations show that in some areas the thicknesses of these ice sheets exceeded two miles. This great volume of ice created loads on the earth's crust so great that it was shoved down into the hot, semiplastic rock that underlies the earth's hard, outer-rock crust. This was analogous to lowering the freeboard of a boat when loaded with heavy fifty-five-gallon drums of water. Further, when sea level was lowered by the withdrawal of seawater to build the ice sheets, the exposed offshore region was again cut both by rivers forming valleys, such as off the southern shelf of New York by the ancient Hudson River, and into U-shaped valley fjords by seaward-migrating glaciers. This erosion formed valleys like those seen in the glacial and coastal regions of Alaska and other northern mountainous regions today. As the ice melted following the end of maximum glaciation, this ice load was relieved, and the depressed rock pressed into the earth's upper mantle rebounded upward—a situation similar to emptying the water from the fifty-five-gallon drums in our hypothetical boat. The rise would be similar to the increased freeboard the boat would experience when unloaded. However, the rising seawater now covering the old coastal plain would also weigh down the margin, thus making a simple rebound from ice removal more complex in nature than in our boat analogy. It would be similar to some of the water from the barrels spilling into the boat, causing it to be lower in the water than if the boat were dry. In some areas of New England, this rebound has exceeded the rise of sea level following the melting of the continental ice sheets. In some coastal areas, the rebound has lifted and exposed sediments originally deposited in marine estuaries and offshore basins 200 to 300 feet above the present sea level.

Further to the south where the loading effects of ice were not as great, the rising sea level has flooded old land valleys and a hilly countryside that originally extended across the shelf, forming an irregular rocky coast with many offshore islands. The bottom sediments are made up of reworked glacial moraines (glacial debris). Thick beds of kelp and marine plants that require hard substrate to anchor their stocks and "holdfasts" could attach themselves to these solid surfaces of exposed rock. This explains the dominance of many plants along the rocky areas of this region.

South from Cape Cod to Nantucket and Long Island Sound, New York, sand from the terminal end of the giant ice sheet covering the New England region was abundant following the rise of sea level. Elongated sand islands, formed by waves and currents as sea level stabilized approximately 4,000 years ago, dominate the coastline. Sand of primarily glacial origin forms the outer banks through eastern Delaware, Maryland, Virginia, and the coast of the Carolinas. Without the abundant sediment and rock material ground away by the ice sheets and glaciers of the past ice ages in the north and the river erosion of the now low mountain chain that runs the length of the east coast from Maine in the north to Georgia in the south, these dominant coastal features would not exist. To the south beyond the influence of glaciers and ice sheets, the large river systems flowing from the eastern U.S. Appalachian mountain chain cut deep bays and formed estuaries during lowered sea level, which were then flooded when it again rose following the last glacier period. Thus, the coastal plain was separated from the offshore banks and the large estuaries became the Delaware and Chesapeake bays. These southern areas have a low relief because they are formed of material that was originally a coastal plain composed of eroded remnants deposited at the base of the east coast mountain chain near old sea levels. This is in contrast to the flooded shoreline hills of the more resistant granite and hard rock coastal regions to the north, which had high relief and which gained further elevation by rebound of the land surface following glacial unloading.

Farther south from Georgia to the Florida peninsula, the bedrock of the region changes from that of an ancient outwash plain derived from the Appalachian Blue Ridge mountain chain to carbonate rocks, which are derived from shell and coral debris deposited in an ancient shallow sea and cemented together to form limestone. This thick limestone platform which extends back in time at least 100 million years is literally cemented on to the southeastern corner of the United States. As the thicknesses increased, the platform sank; however, high productivity of marine organisms kept its upper surface at near sea level throughout this long period of subsequent geological history, forming the Florida peninsula and the Bahama Banks.

The biological community associated with living coral is presently forming the sediments and rock that in many areas shape today's southern offshore and coastline. The limestone platform therefore was well developed prior to the last Ice Age. During periods of low sea level, when ice covered the northern regions, the limestones, which were often broken and had extensive cracks or "joint" systems, were subjected to rainwater. The freshwater from rain does not contain dissolved limestone chemicals and is "aggressive" in its attack of limestone by solution. The limestones were rapidly dissolved when washed by freshwater, more so at joint patterns, or where freshwater stood still at lakes, or in the water tables below the ground surface. Extensive cave systems formed in the subterranean regions of the Florida and Bahamian limestone platform as sea level lowered and rose in its steplike manner following glaciation. At each "still-stand," when terraces and beachrock were formed in the open sea, cavern systems formed inside the limestone. When sea level receded, the less dense freshwater resting on its denser saltwater surface was also lowered, leaving behind open caverns which then began to fill with stalactites, stalagmites, and other features observed by cave explorers. With a rising sea level, these features, which can only form in air, were flooded. Divers and observers from small submersibles have seen stalactites and stalagmites down to 425 feet below present sea level, giving positive evidence deep within the limestone platform of a great lowering of sea level during the past ice ages. The "blue holes" of the Bahamas—the freshwater springs of the Florida outer shelf—and the cave systems now being explored are part of this flooded cave system, remnants of the lowered sea level.

We cannot leave the southern tropical area without some discussion of the mangrove. The mangrove tree in its various forms creates a unique shoreline that must be mentioned because of its ecological and sediment-trapping importance in the tropical region. This unique plant with its several dominant species is extremely important in stabilizing and protecting coastal areas from storm erosion. The mangrove plant has evolved and adapted its life form to growing in the harsh, shallow-water environment of the tropical coastal plains, which undergoes extreme changes in salinity that most other plants cannot tolerate. The four primary species of mangroves form zones of growth which trap sediments moving down rivers and along the coast in their roots. The deeper form, the "red mangrove," has a reddish bark and grows on stalked roots that hold its limbs and leaves above the water surface. The intertwined root system, by trapping sediment, causes the shoreline to advance seaward. At sea level and slightly above, the "black mangrove," with its very small, fingerlike rootlets protruding up through the sediment, further consolidates and traps the sediment in a network of roots and further stabilizes the advancing shore zone. Higher still are the "white mangrove," and finally the "buttonwood," which colonize newly acquired land areas pioneered by the red and black mangroves.

In addition to trapping sediment, the mangrove is an example of

Mother Nature's own sea-to-freshwater conversion plant. The plant forms fresh water for its own metabolic purposes from highly saline seawater by a process known as "reverse osmosis." The energy for this conversion is provided by the sun during photosynthesis. The excess salt forms crystals that can often be seen by the naked eye as shining particles accumulated on the undersides of the vivid green leaves of both the red and black mangroves, and as small crystal cubes when viewed through a microscope or with a magnifying glass. The mangrove forests of the southern coasts not only stabilize the shore, but additionally, the intertwined root system forms a unique environment for the protection of many of the juvenile fish and invertebrate species important to the offshore ecosystem. Lobster, bass, perch, and other species fished for sport and commercial use utilize the mangrove forests for protection and food. For boatmen traveling in tropical waters, this is a shoreline well worth visiting for a closer look.

SUGGESTED READING

An Introduction to Marine Geology, by M. J. Keen (Elmsford, N.Y.: Pergamon Press, 1968).

The Earth Beneath the Sea, revised edition, by Francis P. Shepard (New York: Atheneum, 1968).

Tales of an Old Ocean: Exploring the Deep Sea World of the Geologist and Oceanographer, by Tjeerd Van Andel (New York: W.W. Norton, 1978).

Dr. Robert F. Dill was born in Denver, Colorado, in 1927. He received his B.S., M.S., and Ph.D. in marine geology from the University of Southern California. Dr. Dill joined the National Oceanic and Atmospheric Administration in 1971. In 1975 he moved to St. Croix, U.S. Virgin Islands, to assume directorship of the West Indies Laboratory, Fairleigh Dickinson University. In his thirty-one years of government civil service, Dr. Dill was instrumental in the development of many NOAA programs to establish the first underwater parks and marine reserves, to regulate the use of offshore regions, and to develop the mining of manganese nodules, and was the technical representative of NOAA for the exchange of bottom instrumentation technology between France and the United States. Dr. Dill has retired from NOAA to join the faculty of San Diego State University as a professor of geology. Working in the SDSU Center for Marine Studies, he will develop research programs concerned with marine minerals, submarine canyons, recent tropical (reef) environments, and comparative examples in ancient limestone environments.

WAVES, TIDES, and CURRENTS

Dr. Henry R. Frey

The operation and performance of boats of any size, whether powered by engines, sails, or oars, are influenced strongly by certain oceanographic conditions. Boats of various sizes and hull designs can be affected in different ways; certain designs perform best and are preferred for the prevailing conditions found in local areas, especially the so-called wave climates. In addition to waves, oceanographic conditions that affect boat handling and performance include swell, tides, currents, depth, and water density.

WAVES

Waves are produced by wind through the exchange of momentum from the atmosphere to the sea surface. The nature of a wave field depends on the wind at the sea surface, the water depth, the time duration of the wind, and the fetch—or distance over which the wind stress acts on the sea surface.

In deep water where the lengths of waves—the distances from wave crest to wave crest—are small compared with the water depth, longer waves travel faster than shorter ones; this is why longer waves overtake and pass shorter ones in the deep ocean. In shallow water where wave lengths are greater than water depths, waves of different lengths all travel at the same speed. The speed of so-called shallow-water waves is proportional to the square root of the water depth. Shoaling water reduces wave speed and increases wave height. The extreme case occurs with waves approaching a beach; the forward speed at the crest is greater than that at the trough due to the difference in height above the bottom. At some point, when the crest travels so much faster than the trough, the wave becomes unstable and breaks, producing surf.

Waves large enough to affect boats are not generated instantaneously the moment when the wind begins to blow; they develop over time and distance. To understand the growth of waves, think of a perfectly flat sea surface, and no wind at all—a rare utopia for those prone to seasickness. When the wind starts, waves just a few centimeters long and a few millimeters high, known as **capillary waves,** are produced first. They are called capillary waves because their principal restoring force—the force that tends to flatten the sea surface—is surface tension. Capillary waves can be observed even on puddles as ripples when there is a gust of wind, and they can be observed at sea, traveling on top of much larger waves.

As the wind continues to blow, longer and higher waves, called **gravity waves,** are generated. These are called gravity waves because their restoring force is gravity. Gravity waves within the wind field are referred to simply as waves. Gravity waves that travel beyond the generating wind field are known as swell (more about swell later).

While a **wave field** grows under the influence of wind, waves of different **periods** and **lengths** combine to produce a **spectrum of waves.** The **spectral growth** of waves depends on wind speed and direction. It also depends on the state of the sea just before the wind blows. For a given wind speed, longer durations (time) and fetches (distances) produce longer period waves. A shift in the period of peak wave energy from about three seconds to about twelve seconds can be expected in coastal seas during moderate storms.

SWELL

Relatively long waves can often be seen at times when there is no apparent local wind. Waves generated by faraway storms—up to hundreds of miles away—reach shore as **swell.** There are several major differences between waves and swell. Waves remain within the influence of the generating wind fields, while swell is found far from the wave-generating wind field. The shape of swell is smoother and more regular than the shape of waves; the peaks of swell are more rounded than the sharper, crested waves. Short-period waves are damped more than longer-period swell; thus swell travels farther.

TIDE-PRODUCING FORCES

Tides are also a wave phenomenon; but, unlike wind waves and swell whose periods are counted in seconds, tides consist of principal components whose periods are counted in from half-days to months. Waves and swell are produced by wind, while tides are much longer waves generated by astronomical forces—by the gravitational pull of the moon and sun on the earth.The total tide-producing force has components that depend on the relative positions of the earth, moon, and sun. Although the sun has far greater mass than the moon, forces due to the earth/moon system are generally more than twice as large as those due to the earth/sun system, because the moon is much closer than the sun. For practical purposes in boating, the tide can be considered to "follow" the moon, with its action modified somewhat by the sun.

Strictly speaking, the moon does not rotate around the earth; it rotates around the center of the earth/moon system, which is about three-fourths the distance from the center to the surface of the earth. At the point on the earth's surface closest to the moon, the gravitational force exceeds the centrifugal force to produce a maximum tide-producing force. At the point on the earth's surface directly opposite to the moon, the centrifugal force exceeds the gravitational force, which also produces a maximum tide-producing force. The force is minimum between these points. Thus, in a complete **lunar day** (twenty-four hours, fifty-four minutes), there are two high and two low waters.

The earth, sun, and moon are in alignment twice per lunar month, at full moon and at new moon. When this occurs, the astronomical tide-producing force is maximum and the response is called the **spring tide** (no relation to the spring season of the year). Minimum force occurs when the moon and sun are at right angles, at first quarter and at last quarter; this condition produces a minimum response called the **neap tide.** There are also month-to-month and annual variations; these are caused by the elliptic orbit of the moon around the earth/moon system's center of gravity, and by the elliptic orbit of the earth around the earth/sun system's center of gravity. The closest point in an elliptical orbit is called **perigee,** and the farthest point is called **apogee.** Tidal components related to the ellipticity of orbits are maximum at perigee and minimum at apogee.

Actual tides do not respond purely to the astronomical tide-producing forces. The responses are modified by land masses that tend to restrict, "steer," and reflect tide waves; by the viscosity of water and bottom friction that cause time lags; by varying depths that influence the speed of horizontal motion; and by the size and shape of continental shelves and adjacent basins that respond as coupled systems.

The term **tide** refers to the rise and fall of sea level resulting from astronomical forces. Tide is vertical motion only. The horizontal motion induced by the tide-producing forces is the **tidal current.** Although both tide and tidal current are related theoretically, the effects mentioned in the preceding paragraph cause large differences between vertical and horizontal components—between tides and tidal currents—large enough to deal with them separately.

TIDES

Knowledge of tides is essential for safe boating. The practical applications of tide information are described in numerous publications for mariners. One of the most popular and established references is *Piloting, Seamanship and Small Boat Handling* by Charles F. Chapman; the book is published by Hearst Marine Books, an affiliate of William Morrow & Co., Inc. The National Ocean Service, a component of the National Oceanic and Atmospheric Administration, produces various publications for boaters; these include prediction tables for tides and tidal currents, tidal current charts, and nautical charts. These publications can be ordered from the NOAA, Distribution Branch (N-C-G331), 6501 Lafayette Avenue, Riverdale, Maryland, 20737.

Practical knowledge of tides, especially the times and heights of low and high waters, can be used to pass over a shoal, perhaps out of necessity; to pay out the right amount of rope while anchoring; or to adjust mooring-line lengths when making fast to a pier or wharf overnight. The heights given in tide tables refer to a horizontal plane called a **tidal datum;** the datum for any particular area corresponds to a low-water stage of the tide. Predictions in the tide tables are based on the same datum that is used in producing nautical charts. Because there are three general types of tides (**diurnal, semidiurnal,** and **mixed**), the datums used are also different. The type of tide found along the Atlantic Coast is semidiurnal, with two high waters and two low waters per lunar day (24.84 hours); the tidal datum is at mean low water (MLW). Along the Pacific Coast, the mixed type of tide is found; the mixed type has two highs and two lows, but there is a large inequality in the high waters, the low waters, or both. For the mixed type of tide, there are higher high waters, lower high waters, lower low waters, and higher low waters. The datum for Pacific Coast tides is at mean lower low water (MLLW). The Gulf Coast has a wide variation in the type of tide; the datum is at Gulf Coast Low Water (GCLW). This datum corresponds to either mean lower low water for a location with mixed-type tides and to mean low water for a location with diurnal-type tides.

Tidal heights relative to a datum can be predicted through use of tide tables, but actual water levels may be affected substantially by other influences. Wind, depending on its direction relative to the shoreline, may cause water to pile up, or to depress the sea surface. Where tidal ranges are small, such as along the U.S. Gulf Coast, wind effects may be even greater than tidal effects. Changes in barometric pressure also cause changes in local sea levels; this is sometimes referred to as the **inverse manometer effect.** When barometric pressure falls, the sea level rises, and vice versa. **Anomolous** river flow can cause large changes in both the heights and times of tides in tidal rivers.

CURRENTS

Currents are generally thought of as the horizontal motion of water, and this somewhat simplified concept is perfectly adequate for boating purposes. However, there can be vertical currents that are usually much weaker than horizontal currents, especially in coastal seas where upwelling or downwelling can occur. Vertical currents are important for the transport and mixing of nutrients and pollutants, but not for boat handling.

There are three major components of currents caused by three different types of forces. The current-producing forces are (1) the **astronomical tides,** (2) **wind stresses,** and (3) **water-density differences.** the three major components combine in varying proportions to form the total current.

Tidal currents are the flow of water that accompany the rise and fall of the tide. In many estuaries and along the coastal seas, tidal currents account for most of the energy in the total currents. There are three types of tidal currents: **rotary, reversing,** and **hydraulic.** Rotary currents are found offshore; they often flow with little change in speed, but with gradually changing direction. The most common type of tidal current experienced in small-boat handling is the reversing current found in bays, estuaries, and tidal rivers; this type of current flows more or less in one direction during the flood stage and, generally, in the opposite direction during the ebb stage. The directions of flood and ebb tidal currents are usually not exactly opposite (180°) in direction; they vary in direction according to the shape of the coastline and the bottom. The third type, the hydraulic current, occurs in canals that connect two water bodies; e.g., bays, sounds, or estuaries. The differences in the times and heights of the tide stages of the two bodies produce a flow between them. In simple terms, hydraulic flows run downhill, from high to low local water levels. The Chesapeake and Delaware Canal is hydraulic, as is the Cape Cod Canal which connects Buzzards Bay and Massachusetts Bay. Hydraulic tidal currents in excess of five knots can occur in some canals and narrow passages. Attempting to make headway against such currents may be extremely difficult, even dangerous. A good common-sense solution is to transit as near to the time of slack water as possible.

Because tidal currents are produced by astronomical forces, they can be predicted. As mentioned earlier, the National Ocean Service produces tables of tidal-current predictions and tidal-current charts; the information in these publications is extracted and often appears in various other forms, such as pocket-sized booklets, in local newspapers, and on local television. The tidal-current predictions must be used with caution during times of exceedingly high river flow or sustained strong winds because their effects on currents are not included in the predictions. A common misbelief among neophytes is that there is an exact one-to-one correspondence between tidal currents and tides; i.e., that the times of slack water (slack before ebb and slack before flood) occur when tide levels stand at high water and low water. There are time lags between tidal currents and tides; the time lags differ at different locations. The prediction tables provide information for specific locations, where currents were actually measured and analyzed.

The shape of shorelines and bottom, together with the friction between the flowing water and bottom, can cause major differences over relatively short horizontal distances. This is where local knowledge is extremely important. Examples of small-scale variations are found in eddies that form on the down-current side of breakwaters or peninsulas, in current shears at bends of tidal rivers, and in cross-channel flows that occur at certain stages of the tide due to geographic features. Local knowledge is particularly important when navigating inlets.

The Columbia River Bar is the quintessential dangerous inlet on the U.S. west coast. Waves and swell approaching from the Pacific Ocean are made steeper during the ebb flow, and, concurrently, a cross-channel flow occurs as a result of higher water in Baker Bay, just inside and to the north of the inlet. The U.S. Coast Guard lifeboat station at Cape Disappointment provides excellent lifesaving services; but, nevertheless, lives and boats are lost every year in crossings of the Columbia River Bar. Jones Inlet on southern Long Island is one of many potentially treacherous inlets on the U.S. east coast.

Wind-driven currents are especially important in large estuaries and sounds such as Tampa Bay, Chesapeake Bay, Delaware Bay, San Francisco Bay, Long Island Sound, and Puget Sound. Winds that blow shoreward at an angle with the shoreline produce a longshore current called the **coastal boundary current.** The coastal boundary current along the Texas-Louisiana inner shelf can sometimes make it difficult to maintain course in approaches. A general rule of thumb is that the speed of wind-driven near-surface currents is approximately 2 percent of the wind speed; e.g., a wind speed of 10 knots will result in a wind-driven current component of about 0.2 knots; the direction will be the same as the wind direction.

The third major type of current occurs because of density differences. In the open ocean this type of current is called the **geostrophic current,** and in estuaries it is simply called the estuarine circulation. The driving force occurs when the so-called surfaces of constant density are tilted with respect to surfaces of constant geopotential. This causes water to run "downhill."

The preceding descriptions of waves, tides, and currents that affect marine recreation are meant to be general insights. More detailed information is required for actual applications to safe boating; the appropriate level of detail can be found in references such as Chapman's *Piloting, Seamanship, and Small Boat Handling.*

SUGGESTED READING

Oceanography, by M. Grant Gross (Columbus, Ohio: Charles Merrill Publishers, 1967).

Chapman Piloting, Seamanship, and Small Boat Handling, by E. S. Maloney (New York: Hearst Marine Books, 1983).

Tide Tables and *Tidal Current Tables* (NOAA National Ocean Service, Washington, D.C., 1984).

Dr. Henry R. Frey has twenty-four years of experience in underwater technology and physical oceanography. He is chief of the Estuarine and Ocean Physics branch of the National Ocean Service (NOS). Dr. Frey manages the NOS programs that produce tide tables, tidal current tables, and tidal current charts. Among his achievements at NOS, he was awarded the Department of Commerce Silver Medal for distinguished national service.

Dr. Frey's career includes seven years of corporate research with Uniroyal, Inc.; nine years of research and teaching at New York Univer-

sity and the Polytechnic Institute of New York; a year of consulting; and seven years of service with NOS. He worked with the governments of Jamaica and Thailand to develop their oceanographic programs. His Ph.D. in physical oceanography was awarded by New York University. Dr. Frey is a member of the American Geophysical Union, Marine Technology Society, and Sigma Xi Research Society. He co-authored three books about diving and underwater photography, and fifty-three magazine articles, technical reports, and scientific papers. Much of his leisure time is devoted to Scottish side drumming; he is a member of the Washington Scottish Pipe Band.

WEATHER WISDOM

Eric Sloane

Awareness probably increases with age. In my youth the wonders of weather went by quite unnoticed, but now each sunset is precious. The drumming of rain on the roof, the hypnotic effect of mist and fog, the whistle of wind, and the boiling of thunderclouds have become background to many of the poignant moments of my life.

Having been the first weatherman on TV and having written books about weather, I now find it interesting to view the past and compare modern methods of forecasting with old-time country folklore. Not everyone can be a meteorologist, but it is easy to be weatherwise, and the pleasures of being close to weather are endless.

Early folklore was not all superstition. In the past, people lived by the weather more than we do; even now their predictions are frequently more accurate than those of the professional weatherman. Actually there have been no new weather instruments since early America, and we still use what Benjamin Franklin had in his weather station—the barometer, the hygrometer, the thermometer, and the wind vane. The only new instruments are not really weather instruments but instruments of communication, such as television, the telephone and radio, satellites, and the camera.

Folklore weather was confirmed in the Bible (Matt. 16) when Jesus Christ said, "When it is evening, ye say, It will be fair weather; for the sky is red. And in the morning, it will be foul weather today; for the sky is red and lowering." Few are acquainted with that passage. Here is another (Luke 12:54): "When ye see a cloud rise up out of the west, straightway ye say: There cometh a shower: and so it is." Not superstition at all, for science tells us that the dry dusty air that causes a deep red sunset will flow eastward and reach you by tomorrow, producing fair weather. Storm clouds rising up in the west will (again) flow eastward and produce a shower; so the Luke prediction still holds true.

Morning dew on the deck of your boat, or a heavy night dew on grass, foretells fair weather; a lack of dew always predicts foul weather.

Flies bite more before a storm because lowering pressure releases body odors, while warmth with humidity causes sweat, making the body a target for insects. When flies become bold, expect a shower.

Distant shores will loom and seem closer than usual when mixing sea level atmosphere eliminates the natural haze of good weather. "When the land looms close and clear," says the old sailor, "a full day's rain is mighty near."

"When the stars begin to hide, soon the rain it will betide" is also sailing folklore, true and scientific. Increasing humidity at a high altitude preceding a storm will cause stars to become "fuzzy" and to finally disappear. Watch out when the stars become faint.

The old storm sign of a halo around the sun or moon also has a scientific explanation, for it is the phenomenon of light shining through ice. Cirrus ceilings are lofty ice clouds that flow before a warm air mass; a warm air mass produces a long (but slow) precipitation, followed by mist and humid weather. Moon halo at night, rain tomorrow.

Swallows and bats fly close to the water before a storm, according to sailing lore. The reason for this involves the sensitive ears of swallows and bats that seek insects by sound. Sensitive ears become painful during the lowering of pressure during the advance of rain; flying as close to water as possible therefore increases air pressure and so relieves such irritation.

Hair becomes limp before a storm because human hair is extremely sensitive to humidity. All the first hygrometers (machines for measuring atmospheric moisture) were built around strands of blond human hair!

"Sound traveling far and wide, a stormy day this does betide." This is an ancient sailing proverb that we all are aware of today. The drone of an airplane or the whistle of a distant train will sound different before rain, as if it were heard through a long corridor.

"When the clouds are bright and high, storm and rain is never nigh." Clouds are simply condensed atmospheric moisture, so the drier the air, the higher that condensation occurs. Fine white cumulus clouds with definite flat bottoms are not signs of rain, but evidence of dry fair-weather atmosphere.

Folklore has it that trees show the backs of their leaves before a rainstorm. There is nothing superstitious about that, for tree leaves do grow in normal fair-weather winds, so when opposite or foul-weather winds occur, leaves are upturned and therefore show their backs.

Lightning from anywhere but a western quadrant is from a storm that will not reach you. This is so because thunderstorms almost always flow from west to east. Therefore, lightning from the east, south, or north is from a storm that is moving eastward and therefore will not flow overhead. Southwest or northwest lightning is from a storm that will give just its fringe rain.

Probably the first sign of weather trend is the wind. No matter how promising the sky appears, anything but a western or western-quadrant wind warns of storm. Because wind flows counterclockwise around a low-pressure (storm) area, if you face the wind and point to your side with your right hand, you will always be pointing to the nearest low-pressure area or storm center. During a hurricane, for example, if you face the wind and point to your right, you will always point to a hurricane's eye, and by this method you may follow the course of the storm.

Smoke reacts to weather by rising quickly in good weather and rising slowly (or even curling downward) before bad weather. By watching the smoke from a steamer on the horizon, you can often forecast tomorrow's weather.

It is interesting to note how air pressure holds delicate material in place; for example, soot in the chimney will readily fall when atmospheric pressure lowers, and the smell of swamps and ditches escapes at the same time, so it may be said that when smells become strongest is before a rain.

A drop in air pressure even affects the human body; aches and pains become more noticeable, and the mind becomes accelerated and irritable. Alcoholism is really a blockage of oxygen to the blood cells, so when air pressure goes down (causing a shortage of oxygen), human beings react exactly as if they were drinking alcoholic beverages; some might become sleepy, others disagreeable, others unsteady. To the creative writer or painter, the effect of a cocktail (or an oncoming storm) might excite his imagination and make him more creative.

The woolly bear and the woodchuck's shadow, stepping on an ant, and other such ridiculous weather lore belong to New England's droll humor, but generally speaking, the old folklore has been established from years of logical observation. The logs of sea captains always included weather observations, as did all the daily farm diaries of early farmers. The student of weather might do well to try a full month's observation of wind direction, types of clouds, and the results of each weather phenomenon; it's the sort of thing that might add a lot to your life. The weather and changing skies are with us every day of our lives, yet we have become so used to them that we take them for granted, and we manage to miss one of the greatest of God's gifts. To look at the sky without really seeing it is indeed a pity.

May we all learn to be weatherwise!

SUGGESTED READING

Waves, Wind, and Weather, by Nathaniel Bowditch (New York: David McKay, 1977).

Mariner's Weather, by William P. Crawford (New York: W.W. Norton, 1979).

Weather for the Mariner, by William J. Kotsch (Annapolis, Md.: Naval Institute Press, 1977).
Folklore of American Weather, by Eric Sloane (New York: Hawthorne/Dutton, 1976).

Eric Sloane began his interest in weather nearly half a century ago. His books, Clouds, Air, and Wind, Book of Storms, Eric Sloane's Weather Book, Folklore of American Weather, Look at the Sky, *and various Air Force manuals, along with having been the first weatherman on television, lend authority to his meteorological paintings. The cloud mural in*

the Air and Space Museum in Washington is his largest, being eight stories high and half a block long.

Sloane's habit of simplifying meteorology introduced the newer term "airology" and changes in weather-map symbols. His simplifications made the Beaufort system of wind symbols an almost antique idea. Recently he built a mountaintop studio in New Mexico at an elevation of 8,000 feet, where he is painting the sky and panorama of cloud forms which he hopes to complete by his eightieth birthday.

EXPLORING SHIPWRECKS and PRESERVING ARTIFACTS

Dr. George F. Bass

Columbus' first voyage to the New World in 1492 contributed the first recorded shipwreck to American waters. Since the loss of his *Santa Maria* somewhere off the coast of Hispaniola, thousands of vessels have gone down along our own east coast, sunk by storms, naval battles, fires, reefs, or poor seamanship. Still earlier, American Indian watercraft sank in rivers, lakes, and, almost certainly, along the coast.

After nearly a quarter of a century of archaeological diving on wrecks in Asia, Europe, the Caribbean, and the east coast of the United States, dating between 1600 B.C. and the time of our own Civil War, I am still excited by each visit to a drowned relic. Boaters sometimes can share this excitement, but it requires patience and hard work.

The major problem of diving on wrecks, whether recent or centuries old, is simply finding them. Some of the books in my bibliography list known wrecks, or relate the details of their sinking, but no list of east coast shipwrecks can lead a boater directly to wrecks. Should he reach the general vicinity of his intended site, by using a marked chart or detailed instructions, limited underwater visibility and limited time underwater may prevent him from locating a specific wreck even after days of diving.

The favored wrecks, in fact, are often miles at sea, where visibility is best. I am told that LORAN bearings obtained from another boater usually will not lead you directly to a wreck—a tiny speck in the ocean's vastness—because your LORAN is probably not adjusted exactly to his. Sometimes it is possible to reach the general vicinity with such directions, however, and then to locate the site with a depth recorder.

All the east coast wreck divers I've talked to tell me that the best way of finding a wreck to dive on is to be taken to the wreck by someone—fisherman, charter-boat captain, or diving-club member—who has been there frequently before and can return precisely to the spot.

How do you start? Local dive shops are the best bet. I'm not sure whether the chicken or egg comes first here, for magazines and books listing dive shops are most easily found *in* dive shops. So start with the Yellow Pages.

Fairly recent wrecks, especially merchantmen torpedoed during World War II, and even the German submarines that fired the torpedoes, can still look like ships. A friend recently described to me the well-preserved ladders and companionways in the hold of *John D. Gill,* a tanker torpedoed in 1942 off the coast of North Carolina. I've seen aerial photographs of the shallow wreck of a Civil War blockade-runner off the same coast; the outline, including paddle wheels was amazingly intact.

Much more recent ships have been sunk by the U.S. Coast Guard or Navy to form fishing reefs. In South Carolina, for instance, one can find out about these from the S.C. Wildlife and Marine Resources Department, Fort Johnson, Charleston.

On modern wrecks divers should beware of entanglement with cables, jagged metal, or nets. It goes without saying that no one should dive deeper than his experience and equipment allow.

Older ships, of great interest to history buffs, are harder to locate than modern metal ones. Wooden hulls in the Atlantic are devoured quickly by marine borers, which means that any surviving hull parts lie invisible beneath the sand or mud that protects and preserves them. Only arms, ballast, anchors, or, on more recent wooden vessels, steam-engine parts may present visible clues.

Discovery of an unknown historic shipwreck thus depends either

The raising of the Brown's Ferry wreck.

on incredible luck or on months or years of archival research, followed by a survey with remote-sensing equipment such as sonar and magnetometers. Such sophisticated gadgetry is beyond the interest or means of those with only a casual interest in wreck hunting, but some boaters have persevered and made major discoveries.

It is in the area of discovery that boaters can make their greatest contributions to archaeology.

Most excavated wrecks on the east coast were discovered by amateur divers rather than by professional archaeologists. The American privateer *Defence,* scuttled in Penobscot Bay, Maine, during the Revolutionary War; a British ship of the same period, lying off Cornwallis Cave in the York River, Virginia; and the colonial Brown's Ferry wreck raised from the Black River in South Carolina are prime examples.

In each of these cases, professional archaeologists were alerted, and full-scale excavation followed. Their results are shared with the public through books and articles, and through displays in the Maine State Museum, Yorktown Victory Center, and a planned exhibit in Georgetown, S. C. In each case the discoverer was invited to take part in the excavation.

A boater, presumably, wouldn't want to keep his boat in one place for the years it takes to excavate a wreck—a barge is usually a more suitable excavation platform anyway. But why shouldn't you excavate a shipwreck yourself if you have found it? After all, almost any diver can map, dig, and raise artifacts with care.

What seldom is understood is that fieldwork is only a very small part of archaeology. For every month the archaeologist dives, he works for years afterwards on land. It is after the diving that by far the most time and money are spent on any wreck. The archaeologist has learned to conserve, restore, interpret, and publish his findings, for the benefit of others, by studying for as many years as a medical student. A sport diver who salvages old bottles may be a collector or a history buff, but he is not really an "amateur archaeologist"—we don't hear of "amateur dentists" or "amateur surgeons." Field techniques can be learned in a matter of weeks, but not archaeology.

Let's consider the Brown's Ferry wreck in South Carolina. Its most important artifact is its wooden hull, fastened with iron nails. Wood and metals raised from water will soon disintegrate without expensive chemical or electrical treatments, sometimes lasting years and requiring constant professional supervision. The salvage and restoration of the Brown's Ferry ship eventually will have cost hundreds of thousands of dollars, but the hull and its contents then will be seen and enjoyed by generations of tourists and scholars. Compare this end result with the rotten and forgotten timbers and unrecognizable iron cannons I've seen in the sheds and garages of well-intentioned divers.

In the late 1970s I excavated a medieval shipwreck in Turkey during just three summers of diving. Now, four full years later, I still have a full-time staff of artists, conservators, photographer, archaeologists, and students working year-round on the remains, with a dozen colleagues in this country doing research. The work will continue for another five to ten years, but it is as much a part of the archaeology of the site as the digging. Without it, our excavation would amount to no more than selfish destruction of an irreplaceable cultural resource. With it, millions of people already are learning something new about the past from books, articles, and television programs.

All this explains why anyone who wants permission to excavate a wreck in state waters must demonstrate to the state his expertise, experience, and financial backing—and why the wreck should be excavated at all.

Don't be disappointed if there is not a rush by archaeologists to excavate the Civil War ironclad you've just discovered. If I may use the medical analogy again, a doctor doesn't operate simply because he has a patient in his office. There must be a reason. There is now only one archaeological excavation of a shipwreck on the east coast of the United States (and one in Canada and one in the Caribbean). The excavation in the United States, of one of General Cornwallis' ships near Yorktown, Virginia, is well worth a visit.

By now you may think I've tried to take all the fun out of historic-wreck diving. I promise you that a great deal of pleasure and excitement can come from discovering and identifying historic wrecks.

A New Jersey housewife, with no archival training, tracked down the history of wreckage she found near her beach house. She read old newspapers. She consulted ship registries. She corresponded with authorities. She talked to old-timers. After months of research, she was rewarded with the discovery that she had part of the *George R. Skofield,* built in Brunswick, Maine, in 1885. After more months of hard work, she arranged for the remains to be transported to the Bath Marine Museum in Maine to be part of a larger exhibit on the Skofield shipping family. At the end of an article on her work, Susan Langston wrote, "I feel content. I have preserved a small piece of American history."

The first step in identifying a newly found wreck often is dating the artifacts on it. I see nothing wrong with raising a diagnostic artifact or two for this purpose. Artifact removal from wrecks within states' navigable waters, however, is usually prohibited without prior permission. As laws vary considerably, it would be wise to learn them from the state you are in before accidentally breaking them.

In South Carolina, for example, nothing on the bottom more than fifty years old, including fossils, can be removed without a license. The Institute of Archaeology and Anthropology at the University of South Carolina, Columbia, SC 29208, however, has had excellent relations with sport divers by issuing over a thousand hobby licenses. The license, good for thirteen months, requires each licensee to send a monthly report of his activities to the state. Artifacts found by divers belong to the state for sixty days, with the state having an option of keeping 25 percent; in practice, the state almost always borrows important artifacts for study or display, and returns them all to the divers. Mutual trust has led sport divers in South Carolina to report to state authorities discoveries of dugout canoes, Indian pottery, and even unique colonial vessels that might otherwise have been looted.

Let me repeat that state laws vary. North Carolina claims ownership of any submerged cultural material that has remained unclaimed for more than ten years. In Massachusetts, wrecks more than one hundred years old or worth more than five thousand dollars are protected by law. The law in Virginia is not specific, but an age of seventy-five years has been set tentatively by the Virginia Historic Landmarks Commission for historic wrecks; a permit to explore for historic properties may be obtained from the Virginia Marine Resources Commission in Newport News. Florida laws seem even less specific, with administrative decisions on wrecks made by the Division of Archives, History and Records Management, Department of State, The Capitol, Tallahassee, Fla. 32301. Similar decisions are made in North Carolina by the Division of Archives and History, P.O. Box 58, Kure Beach, NC 28449.

Laws protecting historic shipwrecks in New Hampshire, Rhode Island, Massachusetts, Connecticut, Vermont, Maine, and New York are summarized in *A Sportdiver's Handbook for Historic Shipwrecks*, which may be obtained for one dollar, to cover postage and handling, from Northeast Marine Advisory Council, NEC Administration Building, 15 Garrison Avenue, Durham, NH 03824; the book also gives addresses of pertinent state agencies.

These northeast states have joined in a cooperative effort between sport divers and archaeologists by establishing a clearinghouse at the Peabody Museum, East Indian Square, Salem, MA 01970, both to receive wreck information from divers and to provide guidance and information in return.

Let us now assume that you have obtained an exploration permit—which is not the same as either an archaeological excavation permit or a commercial salvage permit—and that you find a glass bottle on a wreck. How will you identify it and, thereby, perhaps the wreck as well?

Like a professional archaeologist you will work with records you make; you won't want to haul a wet bottle from library to library.

Photograph the bottle, and draw an accurate picture of it. Write a description, including dimensions.

The bottle itself should be left in a container of seawater, which you gradually replace with freshwater over a period of months.After any salts have been leached from the glass, the bottle may be placed in a plastic bag pierced with small holes that will allow the glass to dry slowly, over many more months.

Meanwhile, you are looking in libraries for pictures and descriptions of similar, but already identified, bottles.

The variety of artifacts encountered on wrecks of different ages and nationalities is so vast I will not attempt here a list of books and articles on different types of pewter, ceramics, rigging, clay pipes, ordnance, glassware, and the like. Start with the subject index of a good library.

On wrecks outside the three-mile limit of state waters, even modern ones, the diver must let his conscience guide him about the removal of objects. There is a growing awareness among sport divers that wrecks, like coral reefs, can best be enjoyed by other divers, present and future, if they are not dismantled for the enjoyment of a few.

I've uncovered and raised thousands of priceless artifacts over the years, but every piece, to the smallest fragment, has stayed in the nearest public museum, here or abroad.

It's true that there are thousands of wrecks. But the best preserved, the most interesting, are exactly the ones visited again and again by souvenir hunters.

I began by saying that wreck diving requires patience and hard work. Readers of my articles in *National Geographic* and elsewhere learn of the glamour and adventure inherent in the life of an underwater archaeologist. None of them know that I, too, spend dreary months every year acquiring necessary permits for surveys and excavations, in the United States and abroad. They don't know the discouragement I have felt on surveys when I haven't dived on a single interesting site in more than a month. They don't know how many frustrating years it sometimes takes me to identify a single important artifact from a wreck I've excavated.

The satisfaction, ultimately, is well worth the effort.

I haven't touched here on treasure hunting. I don't think archaeological sites and historic monuments on land, whether King Tut's tomb or humble Indian mounds, should be looted or dismantled for the profit of one or two people. I feel the same about submerged historic sites.

SUGGESTED READING

Skin Diver Magazine publishes many popular articles on wreck diving. Serious archaeology buffs will want to read the *International Journal of Nautical Archaeology,* published quarterly by Academic Press in New York. Proceedings of the annual meetings of the Council of Underwater Archaeology (which are open to the public) are published by Fathom Eight, P.O. Box 8505, San Marino, Cal. 91108. The Institute of Nautical Archaeology (P.O. Drawer AU, College Station, Tex. 77840) prints a quarterly newsletter of its worldwide activities for its members. G. F. Bass's *Archaeology Under Water* (New York: Penguin Books, 1970) describes many of the techniques of survey and excavation; and Keith Muckelroy's *Archaeology Under Water* (New York: McGraw-Hill, 1980) shows the major excavated sites around the world. Robert F. Burgess has written a very readable account of underwater archaeology, past and present, in *Man: 12,000 Years Under the Sea* (New York: Dodd, Mead, 1980). Bruce D. Berman's *Encyclopedia of American Shipwrecks* (Boston: Mariners Press, 1973) lists documented shipwrecks dating from the pre-Revolutionary period to modern times. Keith Huntress's *Checklist of Narratives of Shipwrecks and Disasters at Sea to 1860* (Ames: Iowa State University Press, 1979) is especially useful to those interested in researching wrecks. *Wreck Diving in North Carolina,* by D. C. Regan and V. Worthington, is available from UNC Sea Grant, 105 1911 Building, North Carolina State University, Raleigh, N.C. 27650; *Wreck! the North Carolina Diver's Handbook,* by Jess Harker and Bill Lovin (Marine Graphics, Box 2242, Chapel Hill, N.C. 27414), is especially useful but going out of print. The best introduction to the identification and conservation of artifacts is Mendel Peterson's *History Under the Sea,* published in several editions by the Smithsonian Institution in Washington, D.C.

Dr. George F. Bass, born in Columbia, S.C., in 1932, holds a Ph.D. in classical archaeology from the University of Pennsylvania. He directed the first complete excavation of an ancient shipwreck, off the Turkish coast in 1960, and has since excavated four other ancient wrecks. He is founder and archaeological director of the Institute of Nautical Archaeology, and is Distinguished Professor of Anthropology at Texas A&M University. He has written or edited five books and eighty articles on underwater archaeology.

MEASURING the SEA:
Do-It-Yourself
OCEANOGRAPHIC
EQUIPMENT

Dr. Andreas B. Rechnitzer

Dr. Andreas B. Rechnitzer pioneered the use of scuba and deep submersibles for science, developed innovative equipment for determining underwater explosive effects on marine life, collected and inventoried marine animals in new ways that have become standards for marine ecologists, and involved himself in extracting underwater ship remains from the reefs of tropical seas. During the earlier years of his career he found himself at sea without much more materials for fabricating oceanographic devices than that on board most pleasure craft. Six years of graduate school also meant skimping on expenditures, so the most expedient route to getting oceanographic tools that would suffice was creativity. Accordingly, the devices and ideas described in this chapter hark back to those days. Surprisingly, many of the "primitive" samplers have not changed much in function once they were commercially produced. Jerry-rigging is fun and sometimes a lifesaver. On occasion he found himself overweighted with samples while scuba diving. Unused plastic sample bags filled with discharged scuba air became balloons that made the ascent effortless. On one occasion he was drifting in an arm of the Gulf Stream after surfacing from a long scuba dive. With the tending surface craft just barely visible, he tilted the electronic strobe of his camera rig toward the boat, and it served as a daytime lifesaving beacon for himself and five buddies! Although this chapter stresses simple approaches to oceanographic tool construction, Dr. Rechnitzer has designed some sophisticated devices for sampling to ocean depths of seven miles.

This he did during the four years that he was the scientist-in-charge of the bathyscaph *Trieste*. They were used during the *Trieste* descent to the maximum-known depth in the ocean of 35,800 feet.

The mariner can expand his knowledge about the world beneath his keel through measurements and sampling. The purpose of this chapter is to describe and illustrate how to create some effective, inexpensive, and fun oceanographic instruments for use by the deck-bound boater.

As many boaters enjoy getting into the water, particularly in the Caribbean area, this chapter includes descriptions of instruments and methods that can be used by a snorkel or scuba diver. Diving as a means of acquiring samples and measurements has become a universal scientific method. In some instances it is the only method available to validate measurements, samplings, and observations. The use of instruments by divers is still in a state of evolution, so perhaps your ingenuity may add a new device.

The oceanographic instruments described in this chapter can, in most instances, be improvised from components normally onboard. Rope-line, boat hook or broom handle, empty food container, a man's old sock or lady's pantyhose, a plastic bag, a wine bottle or jug, grappling hook are but a few examples. See a more complete materials list at the end of this chapter. The list is a memory jogger to include items that may not be onboard, but should be if all of the instruments and techniques are to be tried. Most of the materials are readily available from hardware or drug stores. Only the basic features and principles are described, as some ingenuity and creativity is left to the boater as part of the pleasure to be derived when doing-it-yourself is individuality.

SOUNDING AND LOWERING LINE

The basic unit for the surface-bound boater is the lowering line and lowering weight. Properly calibrated with depth units, it can be more eloquently called a sounding line. Since it will be used frequently as a key component, a 100-foot length of ⅜-inch manila hemp rope should be dedicated for oceanography use. Nylon, like most other synthetic fiber rope, tends to strech under a load. If you are looking for accuracy, then stick to hemp rope. The line you are going to use should be fitted with ⅜-inch rope thimbles at each end so that it can be rapidly attached and detached from a variety of devices.

To calibrate the line with durable and meaningful markings, a decision needs to be made whether to use feet, fathoms, or meters. Each is convertible to the other. The scientific standard is the meter that equals 39.37 inches or 3 feet 3⅜ inches. The fathom remains a handy unit since it is built into our arm spread. Being readily available, it has long been the measurement of the fisherman and mariner. The fathom can be approximated by the adult boater by simply spreading the arms out level with the deck—fingertip to fingertip for a 6-foot man usually comes very close to 6 feet without adjustment. Mark 6 feet off on a piece of line. Grasp the line between the thumb and curled forefinger, and stretch the line across the chest. If slack remains, move the arms backward to make it taut (this is often required for people less than 72 inches tall). Remember the position where the fathom was made taut. Repeating that position as the line is paid out or taken in will give a quick measure of the sampling depth. Some like to have a similar fast method for ½ fathom or 3 feet. Your nose is a halfway point. Hold one end at your nose while looking straight ahead. Grasp the end of the line as above and stretch as before, but with only one arm.

The interval you choose to use on your line can be made by painting a bright band around the line. A different color at each tenth interval simplifies counting. For an accurate measure, the line should be loaded with a 15-pound weight when marking. Hemp, unlike the synthetic-fiber ropes, does not stretch significantly, so is the best for this purpose. Less durable markings can be made using electrical tape, string, and other materials.

SOUNDING WEIGHT

The type of line just described when fitted with an elongated lead weight served the earlier mariner as the bottom sounding line. Don't forget it can still fill the bill when the echo sounder isn't functioning. A sounding lead can be purchased from a marine supplier or can be fabricated from a variety of high-density items: a window sash weight, a length of pipe filled with sand or cement, lead weights from a diver's belt, scrap metal, a light anchor, or a pipe wrench. The more streamlined and dense the sounding weight, the more rapidly the line can be paid out, and the less likely it will be to hang up on the bottom. To sample the bottom, the weight tip can be hollowed out and the depression filled with beeswax. When the weight strikes the bottom, a small amount of bottom will become embedded in the wax. Even wheel-bearing grease will perform the recovery function.

WATER CLARITY

Boaters traversing waters, shallow and deep, often wonder how far it is to the bottom or how far down one can really see. A simple device is the Secchi disk. A white disk eight inches in diameter that can be fashioned from plastic, wood, or metal will give the boater his own Secchi disk. The disk must be kept horizontal with the surface. Stability can be obtained by fitting the center with one eyebolt end of a turnbuckle. Remove one of the threaded bolts from a complete turnbuckle. Add a large washer and nut on each side of the disk. This will provide strength and suppress wobble. Leave enough free threads to fit back into the turnbuckle. Lock it with the nut. Now you have a connecting point for the lowering line and the take-down weight. Raising and lowering of the disk at the depth of visual extinction will give a meaningful and repeatable indication of water transparency. Lowering the Secchi disk on the windward side of the boat will keep it in the best position for viewing. Be aware that variations in readings for water having one transparency will occur because of the sea surface changes and particularly the position of the sun.

As a rule, the high transparency common in the open ocean lessens as the coastline is approached. The reasons for this are many and include an increase in plankton populations that thrive on the more abundant nutrients combined with suspended sediment particles derived from storm-disturbed bottom and land runoff. The scuba diver can acquire useful transparency information from the use of the Secchi disk before entering the water to photograph or just to make observations. Here are some typical Secchi disk readings for the east coast: Long Island Sound, 2–3 meters; Woods Hole Harbor, 6 meters; Gulf of Maine, 17 meters; Caribbean Sea, 24 meters.

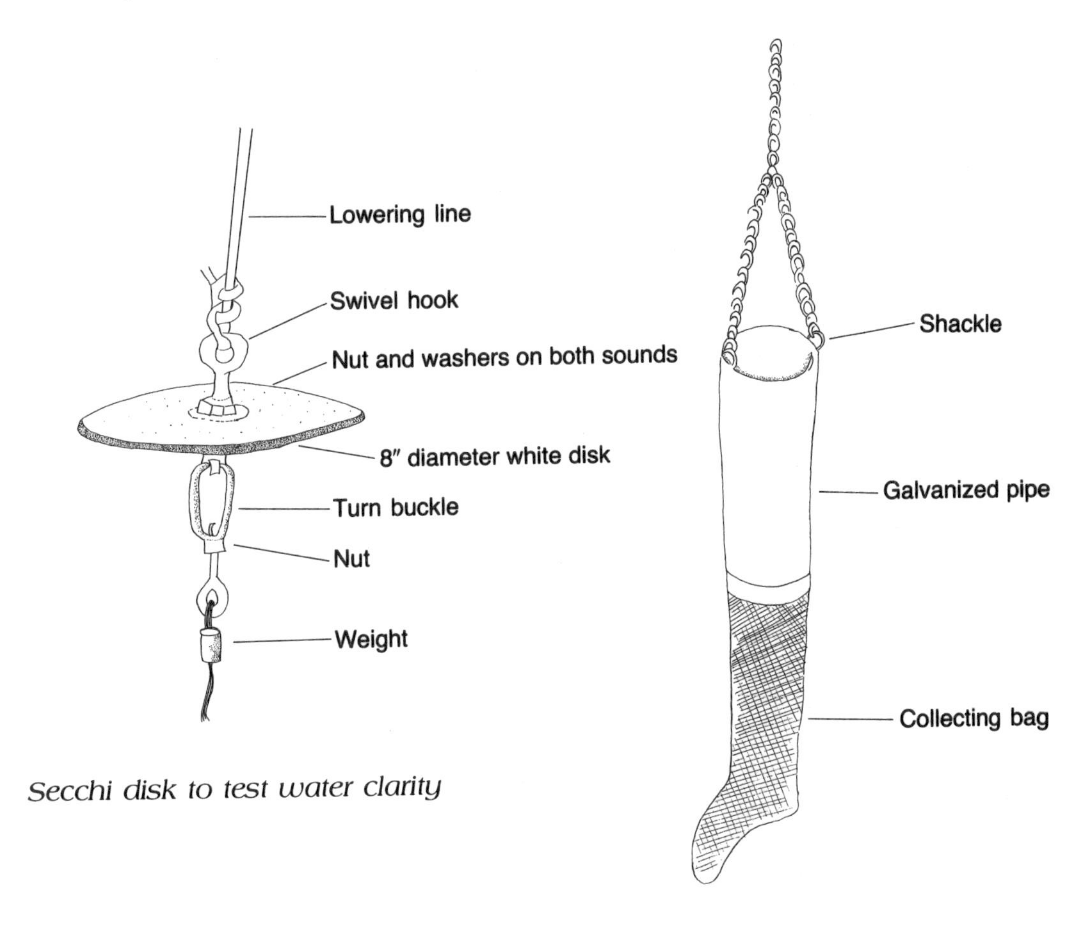

Secchi disk to test water clarity

Bottom sampler

SURFACE CURRENT

The drift bottle continues to be a good means to determine the general direction of surface currents. Sturdy, empty soft-drink and liquor bottles when capped make good drift bottles. Slip a card inside that gives the date, location, your address, and any reward and a request for the finder to fill in information as to place and time of discovery. Toss the bottle over the side and wait for it to be found. The boater has to accept that it may be a long time before he receives a return, but the wait is worth it if he finds out that the bottle traveled hundreds of miles.

Surface current in the immediate area of an anchored boat can be determined by timing the drift rate of orange peels, wood chips, confetti, and any other material that will float. Surface current will often bring an anchored boat's bow into the current. Dropping any of the above items off the bow allows the boater to follow the drifting indicator alongside the hull. Time and distance calculations will give the current velocity.

BOTTOM SAMPLER

To sample the sea-floor sediments and rocks, a pipe dredge can be fabricated easily using a short length of galvanized pipe. One and one-half to two feet of pipe with an inside diameter of two inches or larger is very effective. Add a three-foot-long chain yoke attached by shackles to the front of the pipe 180° apart looking at the pipe end on. To this chain yoke, shackle a single length of chain 3–6 feet long. The combined weight of the chain yoke and the single length will make the pipe dig into the substrate. On the back end of the pipe tie on a man's foot sock. Pull it over the outside of the pipe and secure it with a few turns of high-strength line or a strap clamp. When the lowering line is attached to the single chain, the instrument is ready for use. Let out more line if it comes up empty. More line out makes it dig in more readily.

BOX-TYPE BOTTOM SAMPLER

A larger-capacity box-shaped pipe dredge can be constructed from threaded galvanized pipe parts available from a home-supply store. Band iron, which is also readily available at hardware retailers, is another option. It can be bolted or welded to provide the basic box frame. An eighteen-inch-wide-by-36-inch-long-by-6-inch-high frame fitted with heavy canvas aprons attached to the forward crosspieces (top and bottom) will serve to protect the collecting bag inside the frame, regardless of which side decides to be the bottom.The collecting bag can be a burlap bag or one fabricated from plastic screen, galvanized hardware cloth, or heavy fish-net material; all are suitable. The collecting bag can be joined to the dredge with soft galvanized wire passed through grommets or the bag material. A six-foot yoke of chain and a lot of tow line out will get the dredge to the bottom. Window sash weights added to the tow line about ten feet ahead of the dredge will get the dredge to fish with less line out. Operating any dredge in rocky areas invites a dredge hang-up. Therefore, attach the lowering line to the aft crosspiece. Tie one end of a weaker tow segment to the chain tow bridle and the other end to the tow line, so as to create a slack loop in the main tow line when towing. This arrangement permits breaking the point of tow from the front and shifting it to the rear of the dredge when you get hung up. Usually the dredge will break free of a hang-up using this setup, and a retrieved dredge beats building a replacement.

SITE LOCATION

Site location, survey, and recording will likely arise as needs when you have found something worth returning to again. Marking a site with a buoy is the logical first act. However, remember that untended buoys disappear quickly for a variety of reasons. When in sight of land, an unmarked location can be relocated with no buoy aids, by selecting two or more intersecting range bearings with an angular spread of 60° to 120°. Structures, terrain features, trees, or other fixed terrestrial features offer readily available options.

Record your range-bearing selections in adequate form and keep two or more ranges for each location in the boat. One might be the edge of a prominent building lined up with a conspicuous tree or pole. The second might be a shoreline rock formation that is in line with the notch of the horizon hills. When you are ready to return to your site, line up one or the other range bearings and proceed along a course that keeps the first range-bearing reference in line. When you reach the location where the second range-bearing features are also in line, *you're there.* A third range, if available, should also coincide at the same time.

An inexpensive plastic sextant laid on its side is a very effective method for establishing horizontal angles between prominent features. These angles are used in lieu of convenient range bearings. This use of a sextant is superior to using magnetic-compass-determined bearings.

Despite the comment above, if a location is to be revisited a submerged float anchored at the site will assist relocation, and it is less likely to be heisted. The float can be a plastic gallon jug that is tethered several feet off the bottom or wherever in the water column you select. Paint it bright yellow to make the marker easier to see.

BOTTOM SEDIMENT CORER

A sediment corer can be made from any plastic or metal tube that can withstand the force required to penetrate the bottom. The basic components of a corer are: (1) the coring tube with core keeper (one-half inch to two inches in diameter); (2) a weight of adequate size to provide coring penetration after a short free-fall of the instrument (ten to fifteen pounds); (3) a release mechanism and the lowering line; and (4) trigger. A bail should be installed at the upper end of the coring tube. It will serve as an attachment point for the lowering line. At the top of the bail an eyebolt should be added, the head of which will rest in the quick release. The quick-release mechanism is a modified gate hinge (see

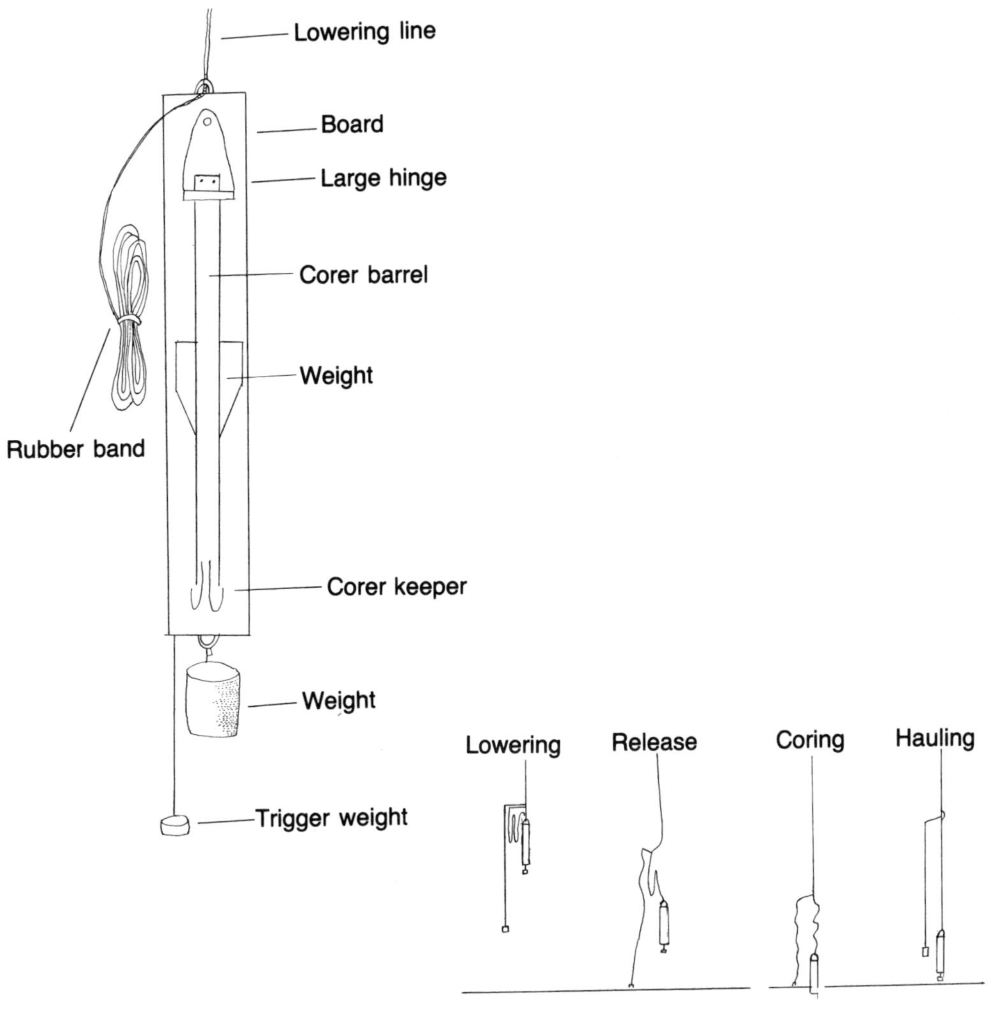

Bottom sediment corer

illustration). A trigger line is added to release the corer for its free fall to the bottom. The weight at the bottom of the trigger line should be approximately the same as the corer.

Mud samples are usually sticky enough to adhere to the walls of small-diameter corers without a core keeper. Sand, on the other hand is very apt to wash out without a retaining device. Oceanographers use an expensive brass spring-leaf core keeper and a check vae at the upper end of the tube. An old sock with the toe removed will suffice as a core keeper when it is installed over the penetrating end of the tube. Slip an inch of sock (tube socks are best) over the pipe, and tie or clamp it on securely. The remaining portion of the sock is then stuffed into the corer barrel. A few trials will indicate how long the sock needs to be for your corer. Cut off what is not needed. Push a rod down as a test. In extracting the corer from the bottom, the captured sediment will tend to fall and bunch up the sock, thus blocking the tube. Greater assurance of a core capture and bunching of the sock will occur when the corer is twisted on removal. To generate a twisting motion, the lowering line should be twisted at your end in the direction that tightens the lay of the line.

A ball check valve at the upper end of larger-diameter coring tubes will help retain a core. The upper end of the tube must be open during penetration to prevent water or air compression. To assure alignment of the ball with the tube opening, a straight guide wire should be driven through the center of the ball. The wire is kept in position by drilling a hole in the rigid bail handle. On the descent the ball check valve must be allowed to open. On retrieval it should automatically fall into position by its own dead weight, water flow, and core "pull." Since this scheme requires extra work on a special bail, consider selecting a pipe size to match the cheap toilet tank stoppers that are available. Select either the stopper with the straight threaded rod that screws into the valve or the all-rubber stopper and hinge that fits over a half-inch pipe.

PLANKTON NET

A professional plankton net for capturing microscopic forms requires very precise fine-mesh material such as standard-grade DuFour bolting silk, used for screening flour. However, an effective plankton net for the boater can be fabricated from a simple open-ended cylindrical container fitted with a yoke or bail. For example, remove the bottom from an empty paint can or bucket and you are halfway finished. Clamp a one- to one-and-a-half-foot-long cone of porous cloth material over the open end (like a section from one leg of a pair of pantyhose). Cut off the pantyhose toe and clamp this open end over a small wide-mouth glass jar. The clamp may be a few wraps of strong line (fishing or linen line) tied off with a bow for easy removal. Disassembly after a tow is not necessary, as the bottle containing water and plankton can be brought up through the cylinder and its contents transferred to another container. Speed for a horizontal tow is usually slow (one to two knots), and the filter rate of the cone should not be exceeded. If a vertical col-

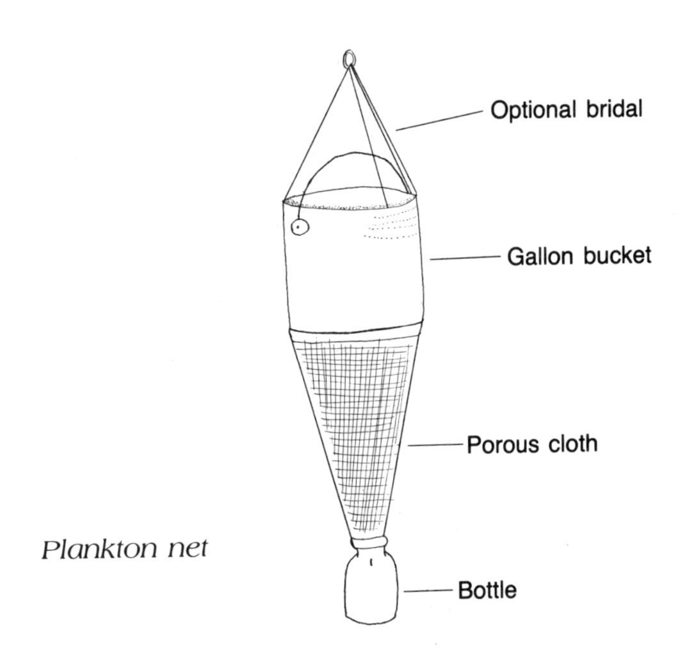

Plankton net

lection is desired, then a small weight should be tied to the bottle end. It will both keep the net oriented properly so it won't fish on the way down and expedite lowering time. Many plankters are visible to the naked eye and a hand lens will reveal a lot of detail. Transfer the critters to a small container so they can be kept in focus.

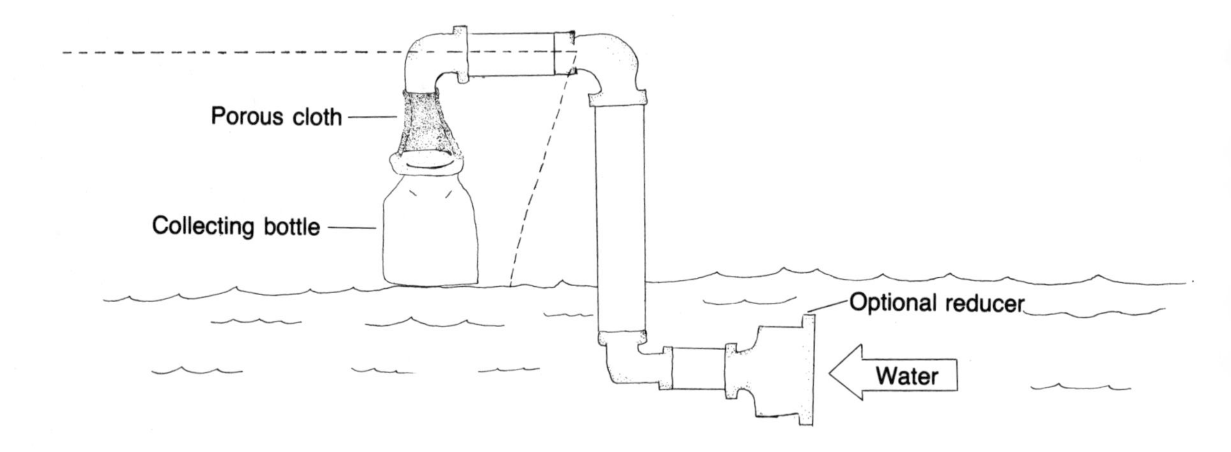

Surface plankton pump

SURFACE PLANKTON "PUMP"

When under way and the lower limb of an "L"-shaped pipe configuration is pointed forward and submerged just below the surface, water pressure will fill it to overflowing, even to deck height. This is then a near-surface plankton-sampler option. Construct the basic unit of one-and-a-half-to-two-inch PVC or galvanized pipe. Adding another "L" limb and a down-directed elbow makes a more convenient point to place a plankton catcher. Any container with a filter barrier around its opening will capture lots of plankton. Again, consider using a segment from one leg of a pantyhose. The finer (smaller) mesh filters will assure retention of the miniscule creatures. The filter can be attached to the down-spout and to a wide-mouth jar in a configuration similar to the towed plankton net. This device can also serve as a steady source of fresh seawater for an aquarium or bait tank. To this unit extra capped pipe can be added to build in stability so that it can be readily hung on the gunnel without special hold-downs and used as a more permanent fixture.

WATER COLUMN SAMPLER (1)

A simple and cheap water-column sampler can be fabricated from a coffee or paint can, a flat rubber sink stopper, and some tape. The top (plastic or metal) of the can is hinged with tape to open and close as a valve. The bottom of the can is perforated from the inside out so a five-inch rubber sink stopper (free-floating or tape-hinged) will cover and

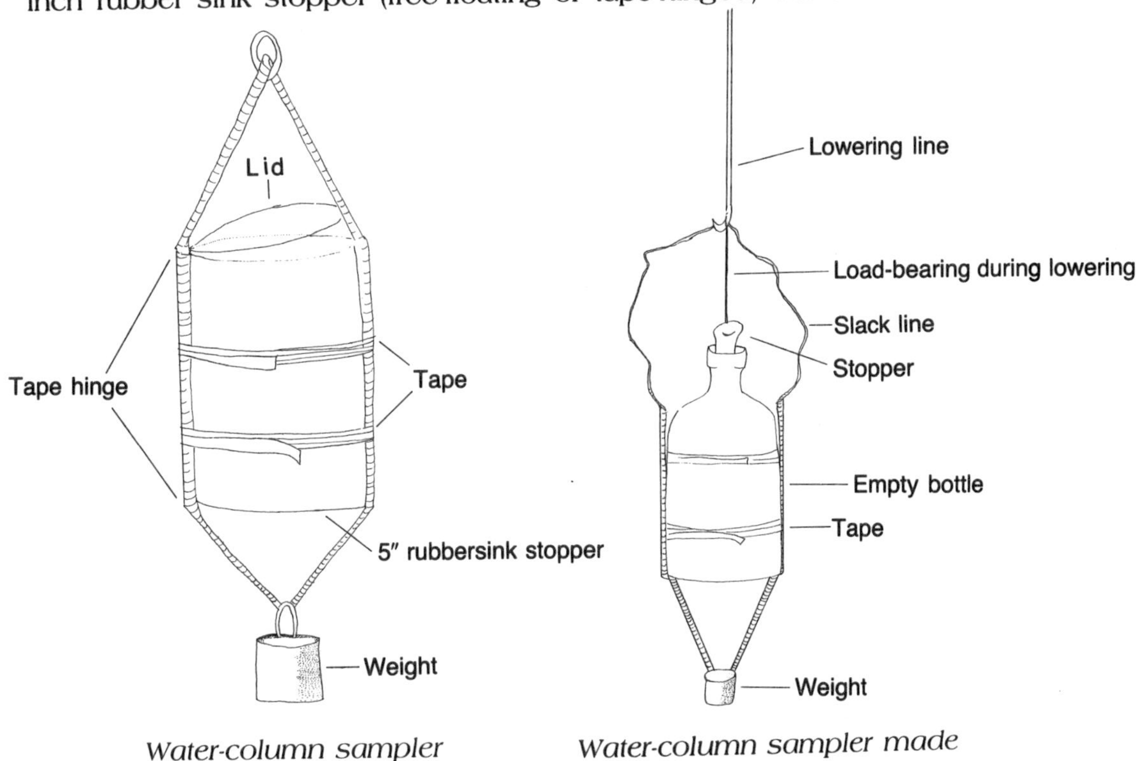

Water-column sampler fashioned from a tin can

Water-column sampler made with a corked bottle

seal the holes when loaded from above the water. The lowering line is attached to the bail. A lowering weight is attached to the can sides by two lines for balance. The lines are taped or tied to the can. To capture a water sample at a specific depth, the sampler is alternately raised and lowered a foot or two to pump water in and out of the sampler to clear it of water captured en route. Once retrieval is started, it should be one continuous motion. This sampler can be used to determine the water temperature at a particular depth. See Water-Column Temperature section.

WATER-COLUMN TEMPERATURE

Use a water-column sampler to carry an inside-mounted alcohol or mercury thermometer that is attached by Velcro fasteners. The water temperature will not change significantly en route to the surface. You can measure the temperature of the water once it is on board and without sending it down, but sending the thermometer to depth provides needed time for the thermometer to reach the correct temperature. Holding the sampler at depth for several minutes before retrieval will give the best results. Read the thermometer while it is still immersed. See next section on Finding the Thermocline.

FINDING THE THERMOCLINE

Most water columns include a level of rapid temperature transition. Usually the transition is from the warm mixed layer near the surface to much colder below. Using the water-column temperature instrument make a series of instrument lowerings. If there is a significant temperature difference between a near-surface sample and a hundred-foot sample, then begin, via trial-and-error lowerings, to find the level of change (the thermocline).

WATER-COLUMN SAMPLER (2)

Here the trick is to send an empty, capped soft-drink or wine bottle to depth and open it there to take in a water sample while using but a single line. The empty bottle sampler is also easy to construct from a half- or even a one-gallon glass jug. Plastic jugs will collapse with the pressure increase, so are not suitable. The narrow-mouth container selected should be plugged with a rubber stopper or a cork. A galvanized eyebolt inserted through the stopper and fitted with a washer and a nut to make it leak-proof when inserted into the jug will perform reliably. A regular cork can be fitted similarly, or a screw-in drawer-pull knob. Two lines, about two feet long, are tied to the eyebolt and then fastened firmly with tape to the sides of the bottle. It is important that a slack loop of equal length (about 4–5 inches) be left in each of these lines before taping. This will allow the cork or stopper to be pulled out after supporting the instrument weight on the way down. A jerk on the lowering line will pull the cork and shift the load to the two lines leading to the take-down weight. The lines should be taped to the bottle at two or more levels to assure that they do not slip when the take-down weight is added to the lines that are joined below the bottle. The ready-to-lower instrument is attached to the lowering line at the eyebolt. At depth the lowering line is given enough of a jerk to pull the plug out of the bottle. Water will rush into the bottle because of the pressure difference (less inside), and the two lines will be supporting the bottle and the lowering weight. A variation of this scheme is to pierce the cork horizontally with a needle or drill and fasten a single line to the cork. The free end is tied to the lowering line to take the lowering load. This sampler will usually perform well to depths of one hundred feet.

The scuba diver can use a similar stoppered bottle and depend on his strength to pop the cork. Caution is advised, as the pressure outside the bottle significantly exceeds the internal pressure (14.7 pounds more every 33 feet of depth). The resulting negative pressure inside the bottle at 100 feet may become a hazard if the opening is covered by any portion of the diver's body, as it will suck on and removal may be painful to impossible at depth. Returning to shallower depths will loosen the grip. Unpredictable implosion (catastrophic crush) of glass bottles can also be expected.

The water sample can be viewed by the naked eye for its usually minuscule inhabitants. A low-power magnifying glass or microscope is most helpful for enjoying the myriad life forms that can be sampled in a one-half-gallon container.

WATER-COLUMN SAMPLER (3)

A first-class water sampler can be fabricated from PVC pipe large enough to be effectively sealed at each end by the crown of a plumber's plunger cup. The large diameter of the pipe allows free flow of water through the sampler on its way down. To make your own you will need, from a plumbing supplier, a piece of PVC pipe and two plumber's plunger cups whose crowns (not concave side) match the pipe to give a satisfactory seal. Stretched surgical tubing, available at most drug stores, will serve as the closing energy source. The two plunger cups are fastened together with the tubing through the PVC pipe section. The center of each plunger cup should be fitted with a small eyebolt for fastening the surgical tubing to the cups. A small thread loop is added to each eyebolt. The instrument is cocked by pulling the two plungers clear of the openings and stretching the surgical tubing until these loops can be hooked to a release mechanism on the side of the pipe. Release of a slip pin passing through the two-thread loop allows closure on command.

One side of a cabinet hinge mounted on the pipe can serve as the receptacle for the release pin. The hinge pin may need to be reduced in diameter with crocus-cloth abrasive paper. As the release pin must be perpendicular to the lowering line, a screw eye should be installed to direct the trigger pull line into the proper angle to free the plunger cups.

A jerk-the-line release scheme is a handy way to close the sampler.

Therefore, two lines need to be added: (1) one to hold the full weight of the sampler and lowering weight (this line is to be purposely parted when the jerk is made); and (2) one to pull the pin free.

TIDE GAUGE

To determine the daily vertical tidal excursion at a strange anchorage, a calibrated vertical staff gauge is needed. A simple two-part instrument involves an anchor and a vertical measure pole that pierces the surface to catch the minimum-maximum of the tide. A metal or wood yardstick attached to a broomstick, boat hook, or longer pole and anchored outside of your swing circle can be read to a high degree of accuracy. The

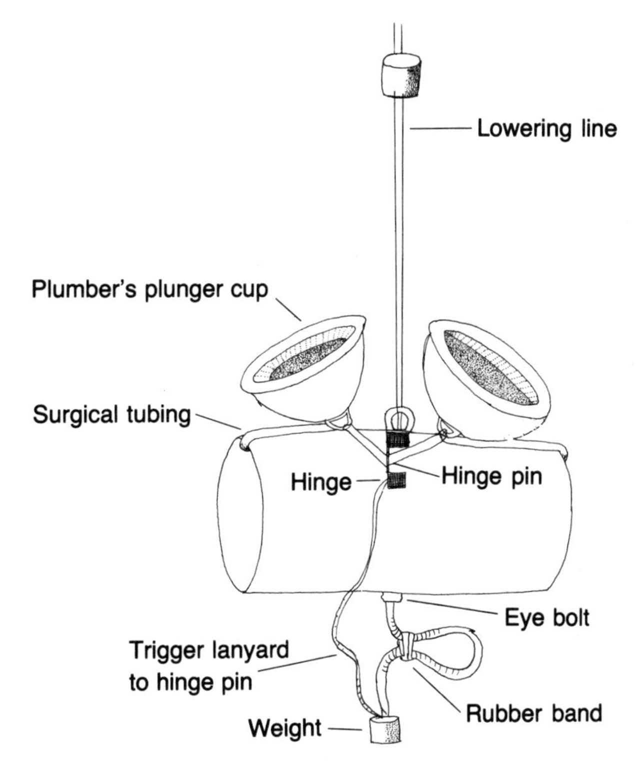

A more complicated water-column sampler using plumber's plunger cups

support pole can be fitted with a screw eye or a drilled hole for attaching the bottom anchor. The pole must be held vertical. Attach an air-filled plastic jug to the staff below the water surface if needed to keep it upright. In lieu of the regular yardstick, the vertical staff can be marked in one-half-foot intervals with painted bands, a wrap of tape, notches, wood screws driven in partway, or nails. These markings are easier to read when the staff is beyond easy reading ranges. These minor changes to the mop/broom handle or boat hook do not alter the staff enough to prevent it from being used again for its original purpose. The same instrument can be used to measure the height of small waves.

Another approach to a tide-gauge design is necessary when waves are present. Small perforations in the bottom of a standpipe allow water to enter the pipe at a rate that is sufficient to match the change in tidal level, and slowly enough not to be influenced by the fast changes in water-depth characteristics of the waves. A straight lightweight wire that pierces the center of a float and moves inside a two-inch or larger standpipe should be dropped into a PVC or other pipe. The wire should extend above the pipe through its full excursion from low to high tide. Read every hour, it will be the indicator of tide with reference to some arbitrary level. A cap on the pipe pierced by a hole will help keep the indicator rod from sticking and tilting.

A variation on the indicator is to hang a small weight over a pulley at the top of the pipe and connect it to the float with a flexible wire or small line. The weight would then move across a prepared scale clamped to the vertical pipe. This instrument needs to be restrained from lateral motion. Therefore, it functions best when fastened securely to a rocky cliff, dock, or piling. A marigram (tide curve) can be developed by hourly readings.

WATER CURRENT

A simple water-current instrument design that will allow water-current measurements at various depths is known as a drogue. A water-current drogue is a submerged plywood or plastic cross that offers enough surface area to assure that the drogue is carried with the water mass and at the same speed. Attached to the submerged cross is a surface float that pierces the surface enough for visual tracking. This component can be affected by surface winds, so attention must be given to keeping it small. Two pieces of ⅜-inch or ½-inch plywood 2 feet by 2 feet fitted together in the form of a cross (like the divider in a cardboard box that provides four compartments) can be achieved by making a cut (⅜ inch or ½ inch) from the center of one edge to the center of each board. Sliding the two pieces together along the slots produces the four wings. Reinforcing "L" brackets (eight) will make it a solid cruciform structure. An eyebolt attached at each end of the drogue at the intersection of the two boards will provide attachment points for the tether line and added weight to *just* sink the drogue. The surface buoy can be made of a variety of materials to which a light vertical staff can be attached, e.g., a wooden yardstick or broom han-

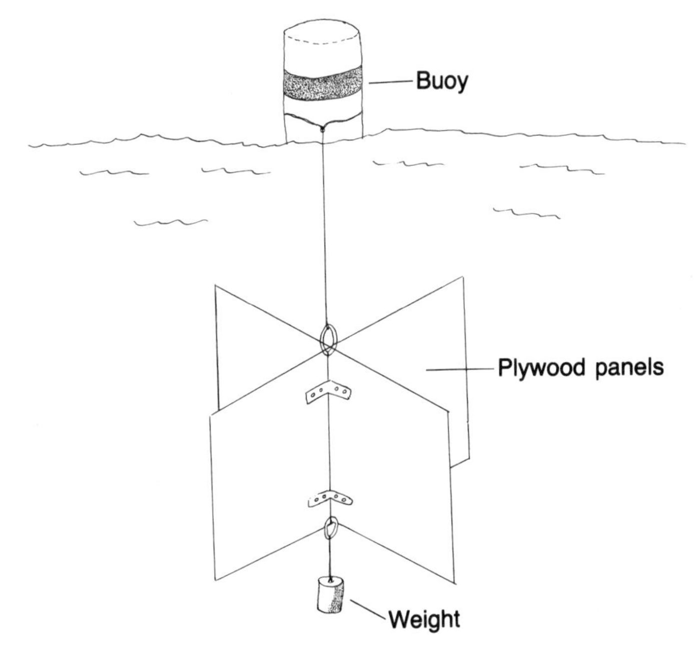

A drogue to measure water current

dle tied to a plastic gallon jug. Ocean currents are now known to run at different velocities and various directions at different depths. The line from the float to the drogue should be adjusted to test various depths.

NIGHT SURFACE COLLECTING

Many small marine organisms and fish, like moths, are attracted to a light source at night. A very enticing light is a regular light bulb (100–150 watts) installed in front of a white painted reflector. Place the reflector about a foot above the water so that the collector's eyes are shaded from the bulb. Use a fine-mesh dip net to capture the variety of small organisms attracted to the light and others that come to feed on the high concentration of food.

A powerful underwater light will perform similarly. Again, the dip net can be used to sample the animals that gather in front of the light. In the event that you want to study the feeding behavior of a basket starfish at night, hold an underwater light close to the basket star. In moments it will actively capture those attracted by the light into the reach of its waving arms. Large organisms such as squid and schooling fish will also seek the lighted area, sometimes in very large numbers.

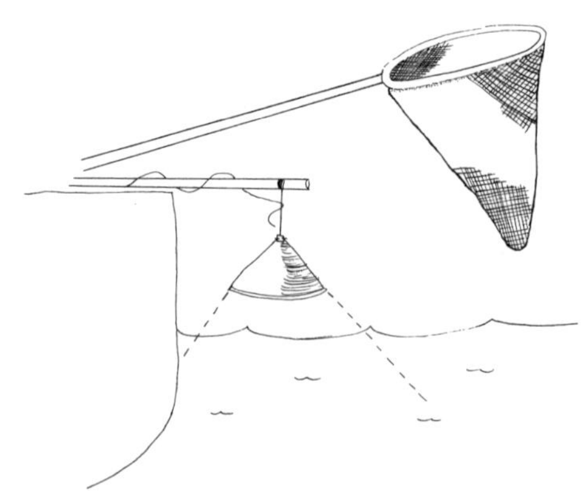

A bright surface light will attract many small organisms and fish at night.

BIOLUMINESCENCE

Bioluminescent organisms are widely distributed in the ocean. Many are microscopic and their individual illumination faint, yet visible to the dark-adapted human eye. The plankton samplers described in this chapter will capture forms capable of producing this living light. Transfer your samples to a jar and seek out a dark location. The organisms frequently need an external stimulus before they will light up. In a jar they will respond to the shaking of the container. The motion of your boat through the plankton population is usually enough. If you have been swimming, those clinging to your bathing suit will glow when struck by freshwater, so turn off the lights and rinse in the dark with *small* applications of freshwater. Stirring the ocean with a paddle or boat hook in an area of high bioluminescence will generate some spectacular fireworks. Photographing the phenomenon requires a high-speed film and longer-than-normal exposure time. Some species have built-in biological clocks that keep them from flashing until nighttime hours are reached. Accordingly, don't be surprised if your daytime collections fail to luminesce—wait until dark.

SOUND RECORDING

Sounds in the sea can be detected and recorded using a tape recorder that can be attached to a microphone or an earphone. To protect an inexpensive earphone made for portable recorders from seawater flooding, simply cover it with a plastic food-storage bag. Better still, coat it with neoprene paint. Your boat's sounds may drown out the ambient noises, so a quiet boat (anchored or engine stopped) is desirable. Living organisms produce noise in almost all geographic locations. Rain is also a big contributor to sound in the sea, so put the microphone over the side when the weather is foul. The microphone need only be immersed a foot to obtain sounds. Deep lowerings are not necessary to hear a majority of the cacophony present.

SOUND SPEED

Sound speed in the sea is approximately five times faster than sound speed in air. The average sound speed is 4,800 feet per second. To measure sound speed, use your underwater microphone to pick up the sound generated by a firecracker floated in a sealed can or sunk below the surface. To extend your range out to a mile, you need the help of a second party to signal when the explosion takes place. A flag signal or CB signals are examples. A stopwatch will suffice for determining the time interval from explosion to arrival. Water depth can be done from your own boat. Signal generation and echo arrival divided by two equal water depth.

DEEP-SEA SAMPLER

Deep-sea sampling can be accomplished in a couple of relatively low-cost ways. A regular marlin fishing pole equipped with a 12V or 24V DC motorized reel filled with piano wire can allow you to comfortably fish and sample to depths of 1,000–2,000–3,000 feet! The small diameter and superior strength of the piano wire is adequate to handle a window sash weight or a large lead sinker and the catch. Some species large and small will take a regular fishook. You'll feel each strike. Other animals are easily lured into a baited trap. A cylindrical hardware cloth trap rolled into a tube and fitted with funnel-shaped entrances is easy to bait and handle. A rectangular trap fitted with conical entrances is another design option. A five-gallon light metal container with a wide mouth lid and punched holes one to one and a half inches in diameter, baited with old fish, will attract and hold some unusual denizens of the deep. These traps can be lowered and retrieved with the electric reel. Lacking the reel and pole, traps can be sent to the bottom with releasable weight and a buoyancy unit. Gasoline is lighter than water, and five gallons will provide approximately ten pounds of lift. The cage and contents (in water) need to be less than this lift force, and should be tested before adding the releasable and expendable weight. An automatic time release can be derived from a few Lifesaver candies that dissolve in seawater. Using such soluble units as a connecting link, the trap will be free for its buoyancy lift back to the surface. You should run time tests on the Lifesavers in 37°–40° water and under tension (equal to the load that will be put on them). Check several flavors for the one that performs best for you. A very lightweight whip should be added so that the free-fall unit can be found when it is again at the surface. Glass fishnet floats can withstand substantial external hydrostatic pressure and yield good buoyancy. A net bag full of these floats is an option to the boater.

DIVER TOW

Towing a snorkel or scuba diver is a fun way to investigate a large area in a hurry. In all instances the towed individual(s) should be required to hang on physically so that on release the diver(s) are free from any possible entanglement.

For towed scuba divers who want to cruise at depth, a weight or depressing force is required. A first method is to use a single tow line with grapnel, tow-line loops, tow bar, boat anchor, or depressor board. In either case, the diver must be able to control his depth by body movements or deflections of the depressor. Towing speed is a slow 1.5 to 2.0 knots. Higher speeds strip off faceplates and put a strain on the teeth and jaws trying to retain the mouthpiece. Observations are degraded also at higher speeds. A 200-foot line ½ inch to ⅝ inch in diameter loaded at the end with 25 pounds of lead, steel, or concrete balls can be used. The divers should hang on at or near the weights. For the convenience of divers who drop off the tow with the expectation of returning to the tow without surfacing, an extra 50 feet of negatively or near-negatively buoyant line should be allowed to trail behind the weights for easy pickup.

A towing board or aquaplane can be created from a flat board or plywood sheet. The size and shape of the board is not particularly critical. Two square feet of surface per diver can give sufficient pressure to raise and sink the towed diver and still not cause excess fatigue. Bridle schemes may vary from a single "V," where a single line is attached to the board by eyebolts or through a drilled hole. Drilling the hole about one-third the way from front to back will make it easier to change the angle of the foil in order to dive or rise. A variation is to attach two lines on each side; one at the front, one at the rear, and the same on the other side. The side lines must be of unequal length, so that the board is level with the direction of tow. To reduce arm strain, add a T-bar "seat" to the back edge of the depressor by a length of rope short enough to allow the diver to sit on it and still move the board. An in-

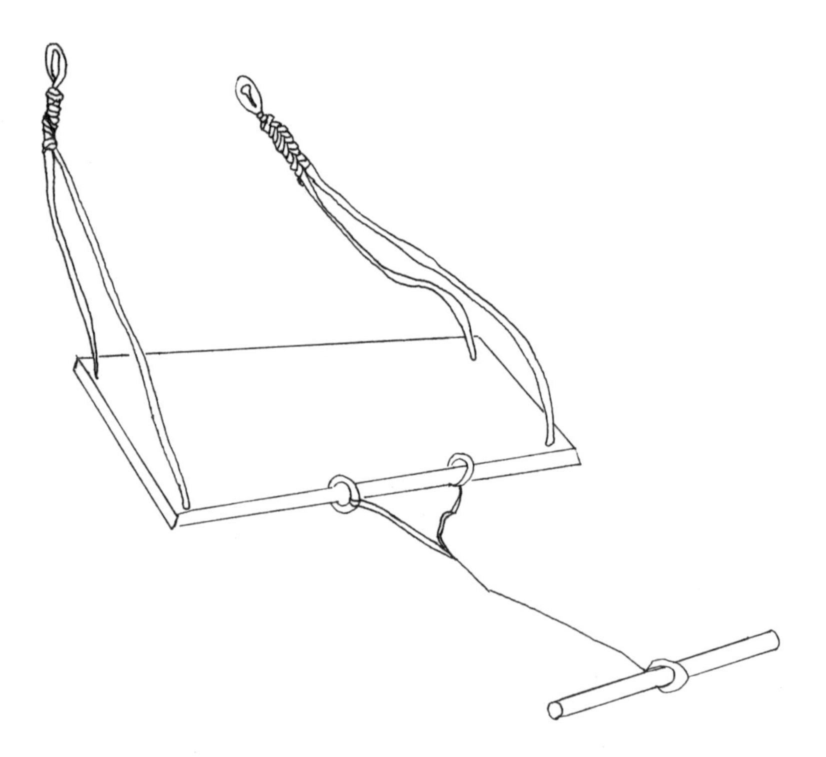

Use a diver tow to investigate a large area quickly.

dication that the divers have dropped off the tow can be accomplished by adding buoyancy to the aquaplane so it will surface. A surface buoy attached to each diver will also show that they are no longer moving with the boat. For safety, smoke and light flares (strobe lights too) are available for those planning to operate in rough water, areas of strong current, or other hazardous conditions.

UNDERWATER RECORD KEEPING

For recording observations underwater, a soft graphic pencil and a white plastic board meet the need. A number of plastic sheets like frosted mylar fit the bill. Ascot papers, Appleton Papers, Inc., also serve the purpose well and are available from most art stores. Several pencils should be available, as sharpening pencils underwater is difficult. Pencil holders can be fashioned with a band of rubber or tethered by a line. Velcro kit glue-ons are handy and readily available in precut pairs from the variety section in most chain drug stores.

To upgrade the slate to a more multipurpose tool, a plastic protractor can be added for measuring angles of sea-floor features such as ripples. An edge can be devoted to linear measurements by adding a plastic rule. Using preglued Velcro components, the list of possible add-ons is wide open to your choice. A brass hook with a spring-loaded clip can be added for easy attachment to a belt.

METER SQUARE

To count organisms on the sea bottom, a meter square has become a professional standard. Fabricating a meter frame light enough for a diver to carry is easily achieved by bending a three-sixteenth-inch rod 90° at one-meter intervals to form the four sides. If a cross made of the same material is fastened by line or wire, the square can be divided into quarter-meters. A quarter-meter may be a more preferable area to study or sample. The frame is laid on the bottom, and sampling or counting is accomplished. If the sampling is to include the inhabitants in the sediment, a four-inch or larger aluminum sugar scoop is very effective; so is a two-pound coffee can, for scooping the area.

A plastic window-screen bag can serve here as both a collecting bag and a sifting screen for larger organisms. Plastic window screen has a number of uses at sea because it is strong, withstands salt air and water, and is easy to fashion into a variety of useful configurations. Keep some on hand.

DIVER CORER

Samples of the substrate and the creatures in it can be sampled using a simple coring tube of pipe. The top of the corer should slip into the mouth of a wide-mouth sample container. The diver can slip a stiff

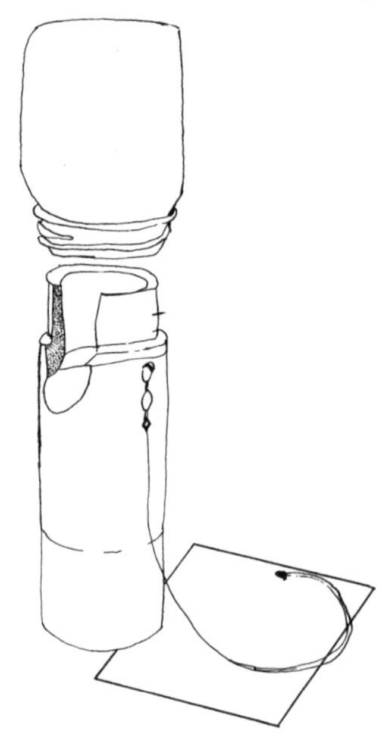

A diver corer is a simple tool to sample the substrate.

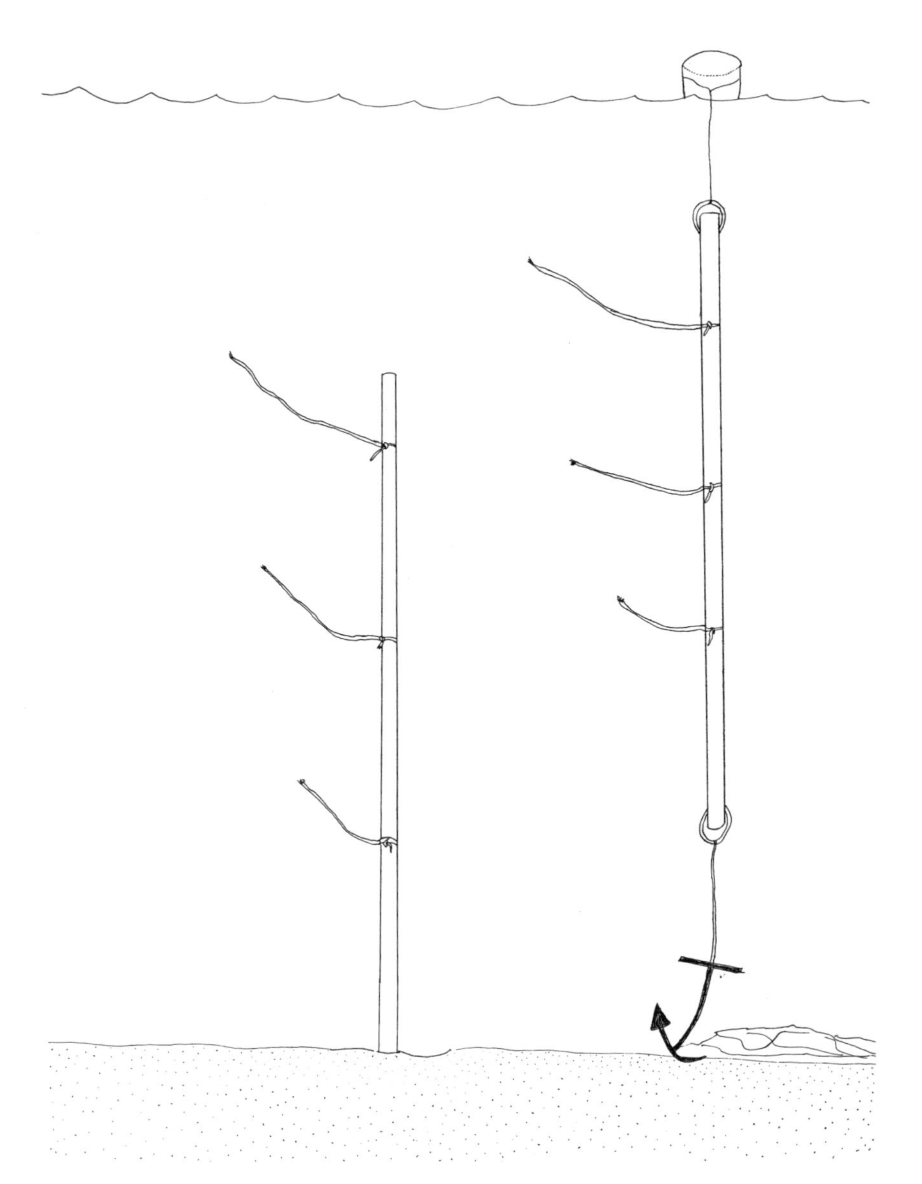

Orbital motion poles

plate of aluminum or plastic under the corer mouth to hold in even a sand core. The diver presses the corer into the substrate about 5–10 cm; he tips it sideways slightly, and then slides the plate into the sediment and over the corer mouth. The corer and plate keeping the core intact are removed simultaneously. The heavy contents of the corer are transferred underwater without spilling because the corer body is inserted below the lip of the jar. Clipped to the corer body, a PVC or galvanized pipe 15–20 cm long suitable for use are readily available. If a collar is desired for a better seal on the sample body, then select two pieces of pipe and join them with a coupling.

UNDERWATER CURRENT MEASUREMENTS

Enjoy the revelations of water motion that can be illuminated by fluorescein dye. A saltwater and fluorescein mixture added to the ocean will flow with and tag water masses. It can be photographed as well as be seen. Current direction and speed can be determined accurately. A point source of dye can be carried to a midwater location or to the bottom in a bottle. Opening the bottle starts the study. To identify the presence of horizontal shear, remove the cap and allow the bottle to sink from just below the surface to the bottom. The open bottle will emit a uniform trail of tagged water. The sink rate and trajectory will usually not be too erratic, so that a nearly vertical trail will be produced. If horizontal shear is present, the contrasting water-mass movements will soon be revealed. Using a lookbox, faceplate, or scuba, the vertical trail of dye will be seen moving with the prevailing current(s) at the same speed and direction. Although there will be some spreading of the dye with time, the observation can be followed with accuracy. Add a small weight to the bottle if improved vertical trajectory is needed.

ORBITAL MOTION

Unbroken waves (or swells) that pass beneath a boater's bow give the visual impression that the water is moving in the direction of travel—

not so. A simple instrument will show that there is orbital motion of the water and that there is no movement of water away from its initial position. Also, the size of the orbit decreases with depth. To see this for yourself, a simple vertical array of neutrally buoyant tethered lengths of wool or nylon yarn suspended in the path of a wave will make it possible to observe (by a snorkeler or scuba diver) the path of motion as the yarn responds to the water motion. The instrument is fabricated from a rigid pole with foot-long flags of yarn taped or tied to the pole. To avoid entanglement, intervals that separate the tips of each when pulled toward one another should be selected. To keep the pole vertical, it needs to be anchored firmly to or into the bottom. The pole need not pierce the surface. The closer the top is to the maximum height of the wave, the greater the orbit diameter. As the orbit diameter diminishes with depth, so the yarn lengths can be made shorter in deeper locations.

SAND TRANSPORT

Sand transport in the surf zone and in deeper water can be traced by marking dried sand samples with fluorescein paint (available from a good paint store). The marked sand is placed in a known location marked by a steel rod. Samples taken at subsequent times and distances from the rod can be viewed under an ultraviolet light to reveal the presence and quantity of the tagged sample. Ultraviolet tubes to fit the fluorescent lamp holders are available from stores selling light bulbs. The direction, rate, and amount of movement can be derived from this simple method. Automatic capture of marked sand grains can be achieved by placing flat-surface plates smeared with Vaseline or grease.

BEACH PROFILE

One of the best oceanographic tools available and ready for the boater's use is the Mark I eyeball. It is still used by professional seamen to estimate the height of waves, evaluate weather trends, estimate water depth, discern transitions in water masses, and semiquantitatively measure natural phenomena. A little knowledge and experience can help the boater avoid safety hazards and capitalize on advantages. For example, boaters intending on a beach landing in an unfamiliar location can "read" a sand shoreline profile while offshore. A steep sand beach, when viewed during a temporary calm period, could reflect a recent history of high waves that could prove disastrous to a small boat going to or returning from such a beach. Even during relative calm periods, the waves off such beaches are often described as plunging and are difficult to pass through in a small boat without a dunking or capsize. A shallow beach profile (one with a gentle slope) is more amenable to landings. Such beaches usually contain fine-grain sand, and the underwater profile is likely to be a nearly flat bottom extending out to deep water. The steep-slope beach is usually coarse sand, cobble, and bedrock.

Beach profiles change rapidly, particularly during storms and heavy wave action. Using two calibrated six-foot staffs, a team of two boaters can measure the change. The first staff is rested on the sand. The second is set six feet shoreward. The second boater sights over his fist as it is slid down the shoreward staff until the horizon is in line with the top of the seaward staff. Read the level on the shoreward staff. Repeat daily or weekly.

BOTTLE FISHING

Small empty glass food jars can be used in a variety of ways for sampling and storage of marine life. A clear eight-ounce or larger jar with a wide mouth and heavy wall is ideal for capturing small marine animals, including some of the reef fish beauties and invertebrates—some of which are noxious and not to be handled. Damsel fish are common to most tropical reefs and are usually amenable to herding by hand into the bottle. In the herding hand, the bottle cap is placed over the bottle once the fish has entered the transparent container.

For Do-It-Yourself Oceanographic Equipment
Materials Checklist

* Bags, burlap and plastic
* Beeswax
* Candy, Lifesavers
* Clamp hose
* Cement or concrete mix
* Coffee can, large
* Chain
* Cork or styrofoam
* Dip net
* Firecrackers
* Fish netting and glass fishnet floats
* Flares, smoke
* Fluorescein crystals
* Grease, wheel-bearing
* Hardware cloth
* Hemp rope, 3/8 inch
* Hinges, small and large
* Lead
* Light shade and bulb
* Magnifying glass
* Piano wire
* Paint, yellow, white, and small can of neoprene; some empty gallons
* Pantyhose
* Plastic sextant
* Plastic sheeting
* Plastic window screen
* Plumber's plunger cups
* Plywood
* PVC pipe, fittings and glue
* Nuts and bolts, 1/4 inch and 3/8 inch
* Nylon, frosted sheet
* Radio earphone

* Rope thimbles, 3/8 inch
* Rubber ball, hard wall or solid
* Rubber bands, thick
* Rubber stoppers
* Shackles
* Steel rod and strapping
* Swivel snap hook
* Surgical tubing, rubber 3/8 inch
* Tape, duct and electrical
* Thermometer, alcohol or mercury
* Toilet tank stopper
* Turnbuckles with extra nuts and washers
* Velcro kit
* Window sash weight
* Yardstick, metal or wood
* Yarn, wool or synthetic

Dr. Andreas B. Rechnitzer received degrees from Michigan State University and the University of California; his Ph.D. is from the Scripps Institution of Oceanography. Dr. Rechnitzer has been involved in a broad range of ocean exploration, scientific investigations, education,and management functions. His contributions in the area of technical innovations and systems development are recognized as pioneering efforts, particularly in the United States capability in deep submergence.

Dr. Rechnitzer's honors include the U.S. Navy Distinguished Civilian Service Award, the Gold Medal Award of the Chicago Geography Society, the Richard Hopper Day Award of the Philadelphia Academy of Sciences, the NOGI Award for Science from the Underwater Society of America, the Gold Medal Award from the Underwater Photographic Society, and the Special Award from the National Capitol Film Festival. He has served as the director of the American Society for Oceanography, president (with Jacques Cousteau) of the World Federation of Underwater Activities, and president of CEDAM International, among many other duties. He holds a former world's diving depth record of 18,150 feet. Dr. Rechnitzer is currently the technical adviser to the Naval Oceanography Division, Department of the Navy.

PROJECTS and HOBBIES for the MARINER

Karen Hensel

and

John J. Hensel

Unpacking the mysteries and marvels of the aquatic realm is so often a fun activity. Familiar objects can be viewed, collected, processed, studied, and enjoyed in ways that will teach, give you pleasure, and encourage further investigation at the level of your choice. All projects that have been selected require a minimum of equipment and setup and cleanup time, finished products may be stored in minimal space and will prove to be potentially more exciting ashore—as gifts, collections, conversation pieces, and the basis for a continuing interest in the biology and behavior of the collected plants and animals.

FISH PRINTS/PRESSED SEAWEEDS

A printed or pressed record of trip finds and catches can provide a fascinating vehicle for future decorating and investigations of the marine environment.

Both fish prints and pressed seaweeds can be used in a variety of ways when back ashore—as a private collection or as a very special present. Seaweed prints make terrific note and greeting cards; framed, they are a unique conversation piece and a reminder of harbors and

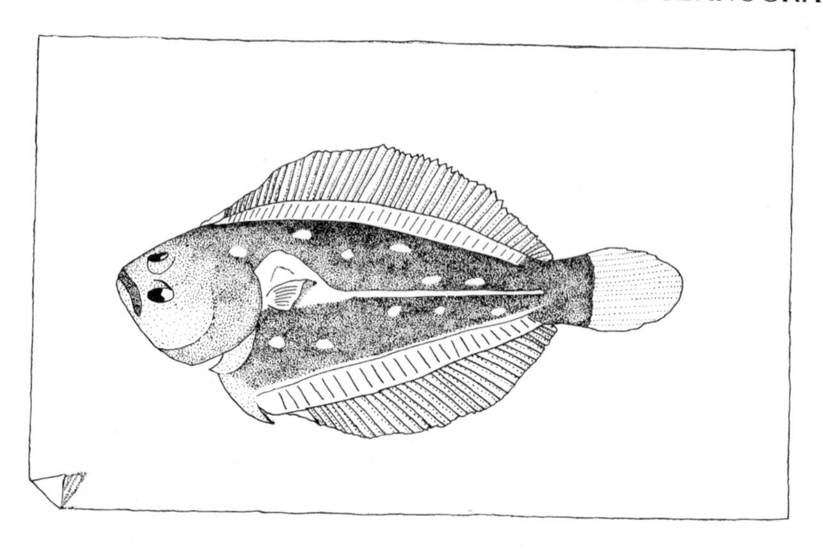

A

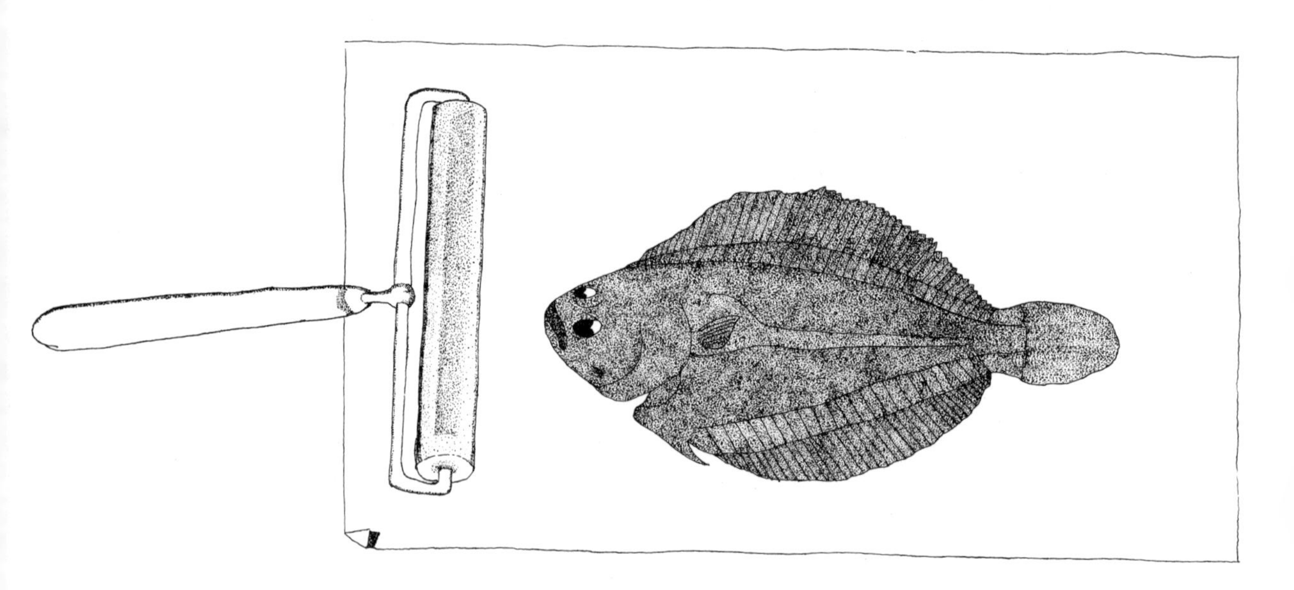

B

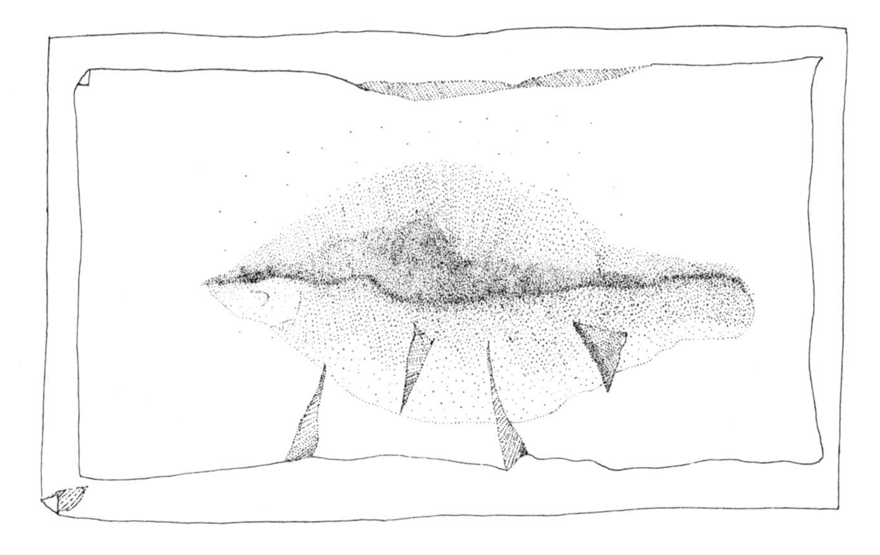

C

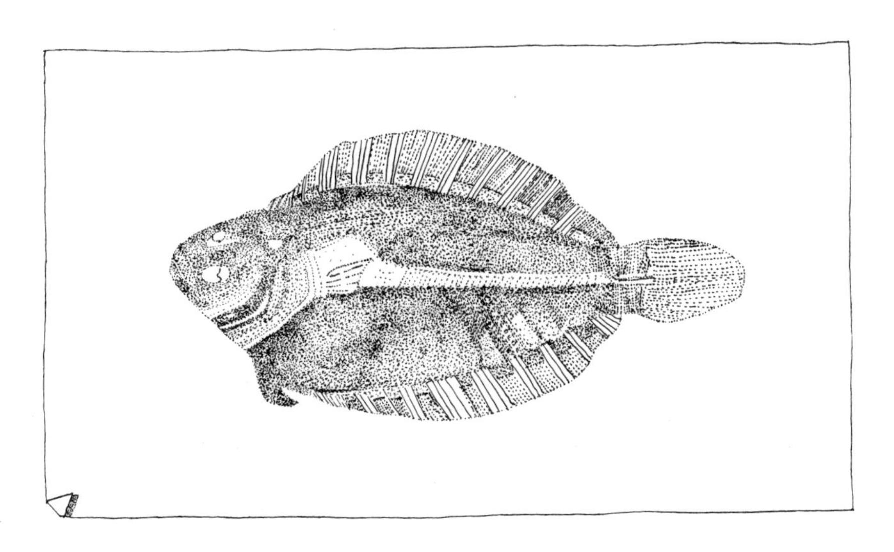

D

inlets explored. As a scientific record, keying or identification of seaweeds and fish, etc., can be both fun and challenging. Oil-base paints used on fabrics will provide an enduring, washable, and useful pattern for pullovers, shirts, pocketbooks, wallhangings, etc. This technique may also be used to make prints of leaves, shells, rocks, and corals.

JAPANESE FISH PRINTING (GYOTAKU)

The catch of the day can be recorded for posterity in a method that has been used for over a hundred years by the Japanese to record and study fish catches, and more recently by American scientists.

Materials:

fish
small roller or brush
paper: newspaper, rice paper, fabric
ink: water-base ink (linoleum-block ink best), oil-base for permanent
    fabric print (optional)
pins, modeling clay.

Directions:

Wash the protective slimy mucus covering off the fish and blot completely dry. Always wash and wipe from head to tail to avoid dislodging overlapping scales. Place fish on newspaper and spread fins (normally tucked close to the body for streamlining). Spread fins out over bits of clay and pin them in position, working from head to tail. Roll or brush an even coat of ink onto the thoroughly dry fish. Place newspaper, rice paper, or fabric over fish and pat gently, gently pressing paper over total surface of fish. Gently lift paper and allow to dry. *Gyotaku!*

SEAWEED PRINTS

At the beginning of the food chain, microscopic algae supports all manner of aquatic animals. Aside from being an annoyance on the bottom of your hull, seaweeds, upon closer inspection, can be as decorative, beautiful, and intriguing as land plants. Scraped from your hull, caught in your anchor line, or collected from the water, unappetizing globs of seaweed can be floated in a container, mounted on paper, pressed, dried, framed, and studied at a later date. Seaweeds contain gelatinous substances, which make them commercially important and interesting, and which we use or eat every day! Your toothpaste probably contains seaweed derivative; your medicines rarely say "shake well" because of them; they thicken puddings, ice creams, soups, jellies, salad dressings, and beer! Materials for this activity take up little space and are readily accessible ashore for the cruiser. A newspaper, several pieces of wax paper, and sheets of porous paper such as blank newsprint paper or rice paper are all that you need to bring you countless hours of fascination.

PRESSED SEAWEED

MATERIALS:

brick or other heavy object
newspaper
wax paper
basin of water (baking pan will do)
construction paper
seaweed*

---

*May be frozen and thawed and reused later.

Directions:

Float seaweed in basin of water. Slip mounting paper under plant—arrange seaweed on paper while submerged. Place one hand under paper and tilt. Slowly lift from water, allowing water to run off. Place piece of wax paper over plant side of paper (so it won't stick) and press between newspaper to dry. Place flat and under weight to dry. When dry, gently peel off wax paper.

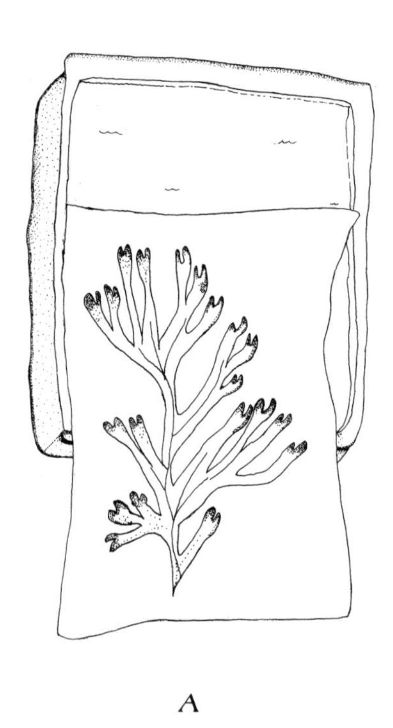

A

B

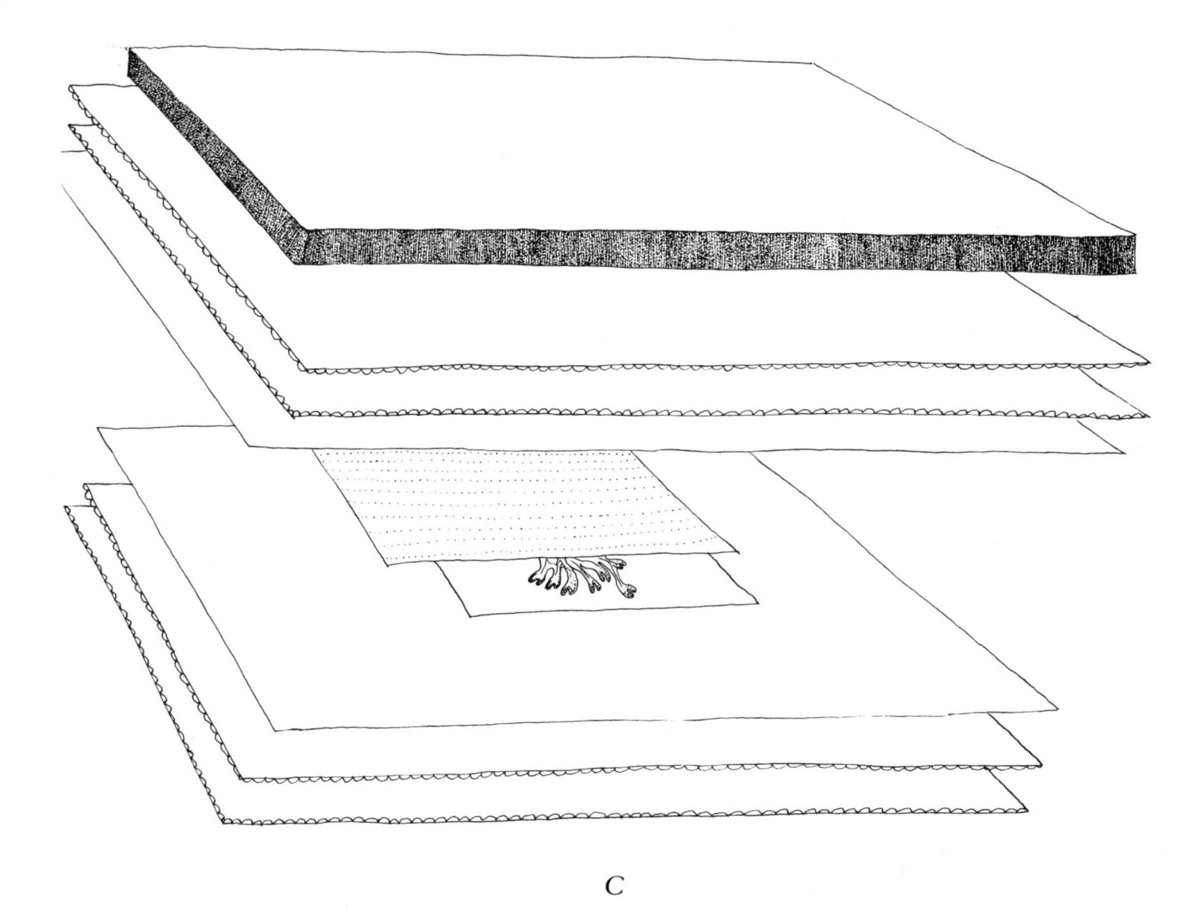

C

D

Scrimshaw—(Old English word for hard work! Dutch for *scrimshander* —meaning "idle fellow.")

Scrimshaw, the American art form developed by 19th century whalers, consists of inscribing a picture or pattern on bone or ivory and staining with ink or lampblack.

The modern-day sailor can while away hours at sea with the barest of essential equipment, etching or carving images in animals bones, plastic, or clamshells instead of whale teeth.

Materials:

sharpened nail
india ink or oil-lamp carbon
cleaned and boiled clamshell, beefbone, or plastic.

Directions:

Thoroughly clean material to be etched; boiling removes all organic material (beef bone may also be bleached to lighten). Sketch nautical or other scene on bone or shell. Etch, using scissors, nail, knife, or other sharp object. Rub india ink or lamp carbon into etched lines for color, then clean surface with tissue.

FISH SOCK OR KITE

When caught in the doldrums, looking for a diversion for the nonfisherman aboard, or an unusual record of fish caught, try creating a Japanese fish sock or kite. Displayed singly or in clusters, the one that didn't get away can provide you or your child with a wind indicator or a charming and very personal toy.

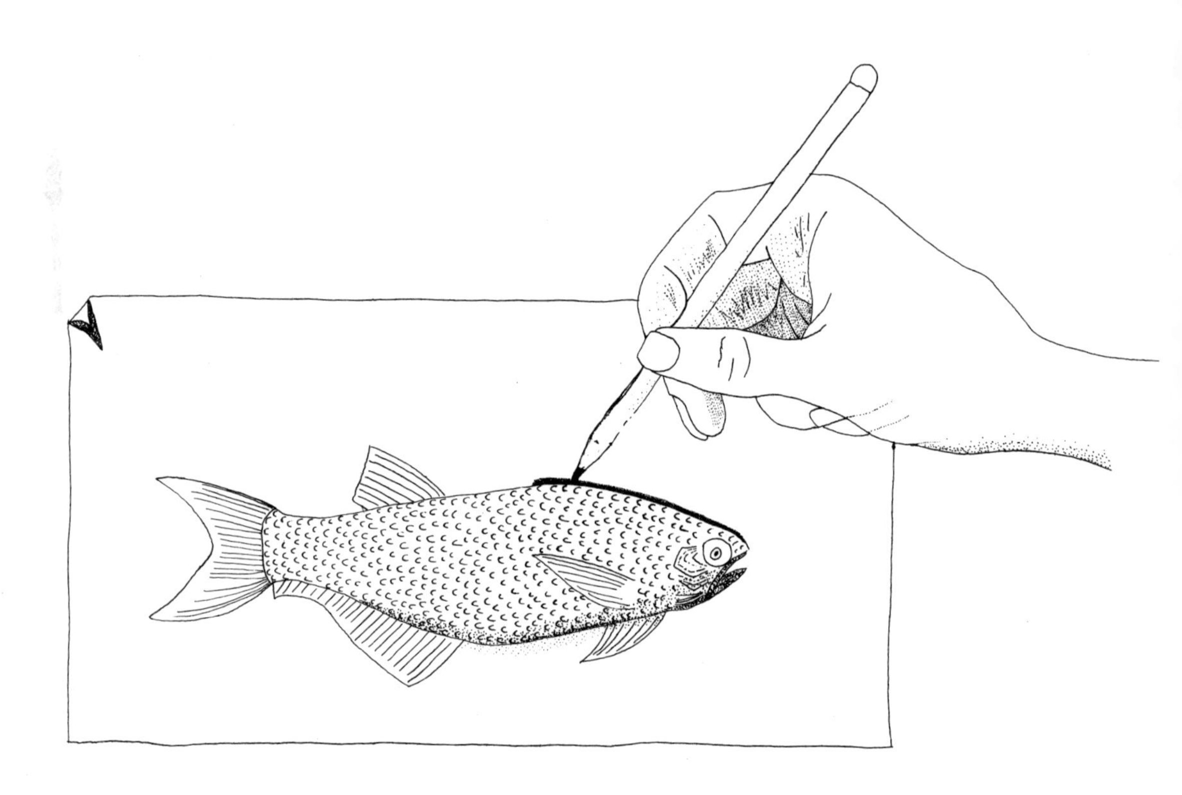

Carefully trace outline of fish.

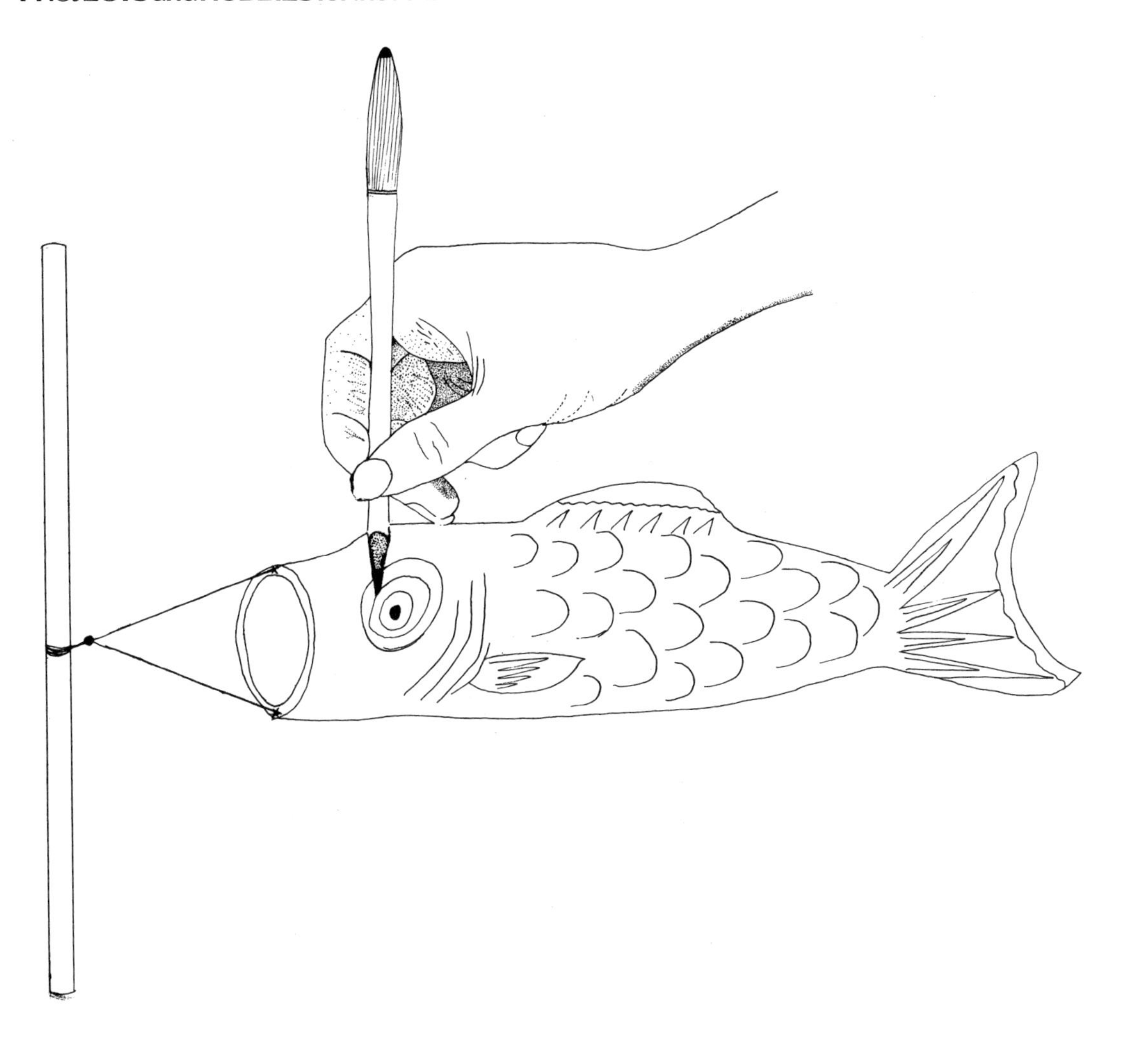

Color fish outline with Magic Markers, paint, or crayon.

Materials:

scissors
needle/thread
string
white heavy-duty fabric or light canvas
colored magic markers/crayons/or oil-base
paints and brushes
ruler
wire hoop—approximately 4 inches in diameter
(or size of open fish mouth desired)
newspaper or tracing paper
fish (the catch of the day that you wish to reproduce)

Directions:

Place fish on newspaper and trace outline carefully, noting and marking position of eyes, mouth, gills, fins, and color patterns. Remove fish from paper; allow paper to dry and cut out fish outline which you will use as a pattern. Measure fish length and height in rectangles two inches larger and wider than fish measurements. Place pattern on each piece of fabric and fish outline (including eyes, fins, gills, etc.). Place fresh sheet of newspaper under each piece of fabric and color fish outline with magic markers, paint, or crayons. When fabric is dry, the two pieces are sewn right sides together along the top, bottom, and back sides, leaving the mouth open. Carefully trim excess fabric or selvage and turn right side out. Stitch metal hoop onto mouth opening securely. Securely attach string to hoop in fish. You are now ready to fly your fish sock from a halyard or shroud.

SANDCANDLES

Sand can be comprised of small bits of quartz, feldspar, garnet granite, and coral, depending upon the local origin of the sand composition. An interesting record of beaches visited can be created by making candles ashore, creating molds out of depressions in the beach sand—sand will adhere to the outside of your candle.

Materials:

small shovel or spoons
pot holders
candle paraffin
candlewicks/stick
crayons (for color desired)
old saucepan or coffee can (for melting wax)
Hibachi or grill

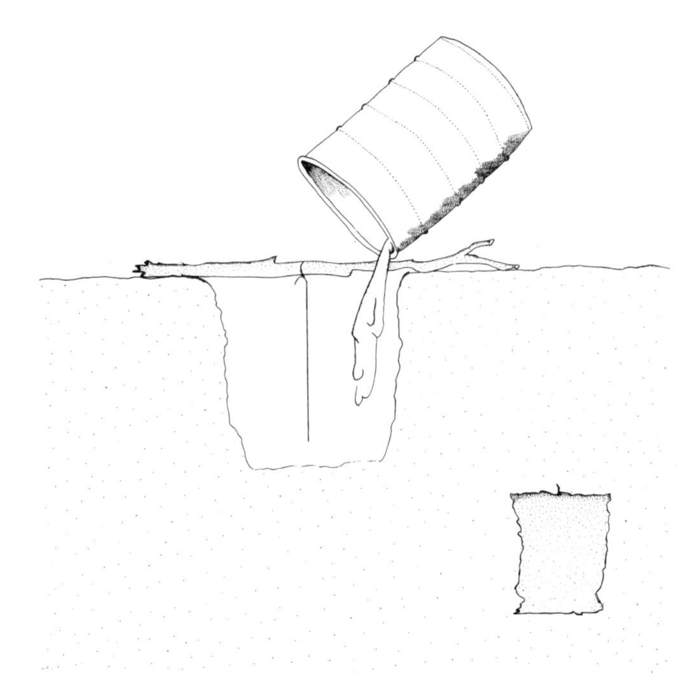

Pour melted paraffin into hole with wick in place.

Directions:

Locate a section of clean, moist sand that you wish to use for your candle mold(s). Set up and start grill nearby. Break up and melt paraffin in coffee can, making sure that wax does not come in contact with direct fire or flame. If a colored candle is desired, add broken crayon bits until desired color is reached. Scoop out depressions in moist sand, the depth, width, and shape you desire candle. Tie one end of wick around a small stick, placing stick across top of mold hole and lowering wick into mold. Pour melted paraffin carefully into hole and allow two to three hours to cool. Later, carefully dig out candle and allow to set for twenty-four hours.

PHOTOS WITHOUT A CAMERA

Another record of beaches visited and objects found can be made by creating impressions of them on light-sensitive paper and identifying them with the aid of a nature guide. Using this process, objects found can be left on the beach and identification can be made when back aboard or under way. Light-sensitive paper will darken where light strikes it. Where objects are placed, it will remain white, giving you an exact outline record of your finds!

Materials:

blueprint paper*
sunlight
cold water
newspaper

*Keep blueprint paper in a dry, dark place until ready to use.

After exposing to sunlight, place blueprint paper in cold water.

Directions:

Collect objects you wish to record. Place blue side of paper up and arrange objects on it. Expose to direct sunlight for one or two minutes. Remove objects and place blueprint paper in cold water for a minute (one tablespoon of hydrogen peroxide in one cup of water will act as a fixer if you wish to do this on your boat—place in this solution first for several seconds, then place in water for a minute). Remove print from water, blot with paper towel, and press dry between newspapers if desired. Print is now ready for framing.

Light-sensitive paper will darken where light strikes it.

A record can be made of beach finds. Later, use the record to identify your finds with the aid of a nature guide, or keep as a souvenir of a day at the beach.

Karen Hensel began her career with the New York Aquarium when she founded the Education department in 1970. A specialist in multigenerational science with degrees from Columbia University, Ms. Hensel has conducted ethnographic research with Aquarium visitors resulting in a new multigenerational facility, Discovery Cove.

A founder of both the New York State and National Marine Education associations and adviser to marine educators internationally, Ms. Hensel believes that marine literacy encompasses the arts as well as the sciences and endeavors to reach broad audiences with marine messages.